Flotation Chemistry

浮选化学

Nie Guanghua
聂光华
Deng Zhengbin
邓政斌
编著

·北京·
· Beijing ·

内容简介

Flotation Chemistry（《浮选化学》）是全英文教材，主要内容包含矿物加工工程专业学生必须掌握的浮选英文基本知识，以及英语写作与表达的基本技能。

浮选体系中，酸、碱等化学药剂的添加及矿物溶解等，使矿浆溶液中发生大量的化学反应，而这些化学反应在浮选过程中对浮选起到至关重要的作用。因此，本书在介绍浮选基本理论的基础上，详细介绍浮选化学基础，之后分别介绍天然可浮性矿物、硫化物矿物、不能溶解的氧化物和硅酸盐矿物、微溶盐和可溶解盐矿物的浮选化学。前言下方附二维码，供学生下载专业词汇，方便学习。

本书的主要特点是理论与实践相结合，包含从基础理论到实际应用的内容，可供矿物加工工程相关专业本科生及研究生，以及从事相关专业的工程技术人员参考使用。

图书在版编目（CIP）数据

浮选化学= Flotation Chemistry：英文/聂光华，邓政斌编著.—北京：化学工业出版社，2023.8

ISBN 978-7-122-43395-4

Ⅰ. ①浮…　Ⅱ. ①聂…②邓…　Ⅲ. ①浮选法-教材-英文　Ⅳ. ①O658.6

中国国家版本馆 CIP 数据核字（2023）第 075035 号

责任编辑：袁海燕
文字编辑：曹　敏
责任校对：刘　一
装帧设计：王晓宇

出版发行：化学工业出版社
（北京市东城区青年湖南街 13 号　邮政编码 100011）
印　　装：北京印刷集团有限责任公司
787mm×1092mm　1/16　印张 12　字数 280 千字
2023 年 8 月北京第 1 版第 1 次印刷

购书咨询：010-64518888
售后服务：010-64518899
网　　址：http://www.cip.com.cn
凡购买本书，如有缺损质量问题，本社销售中心负责调换。

定　　价：98.00 元

Preface

Due to the addition of chemical reagents and the dissolution of minerals, there are a large number of chemical reactions in the flotation system, which play an important role in flotation. Chemical factors include tensions of the air-solid and solid-liquid interfaces, bubble attachment processes, chemical interactions in solution, size and shape of solids, size of all bubbles, and the hydrodynamics of the system. The chemistry of the system, both in the bulk solution and at the interfaces, is most important, since these flotation chemical factors control adsorption and, hence, the selective separation of one mineral from another.

Therefore, this book briefly introduces the basic theory of flotation, introduces the solution chemistry of flotation reagents and mineral dissolution chemistry in detail, and divides them into natural floatability mineral, sulfide, insoluble oxide and silicate, semisoluble salt, soluble salt and fine-grained minerals according to their own structure, and common and different characteristics of flotation. This book discusses the flotation and flotation chemistry according to the above mineral classification.

The objective is twofold: it is intended that the book will be used as a textbook by upper division students as well as a reference book by graduate students, practicing flotation engineers and scientists.

Sincere thanks are due to Ministry of National Education for financial support, to colleagues from the department of mineral processing engineering, Guizhou University for their help. A special gratitude is extended to graduate students Zhu Zhixiong, Wang Zhenggang, Tan Dongdong and Li Jiaxin for their assistance in the editing of some figures in this book.

words

Nie Guanghua

Deng Zhengbin

October, 2022

Contents

Chapter 1
Introduction

1.1 Flotation

Flotation: separation utilizing the different surface properties of the minerals. It is one of the most important methods of concentration, which is effected by the attachment of the mineral particles to air bubbles within the agitated pulp. By adjusting the “climate” of the pulp by various reagents, it is possible to make the valuable minerals air-avid (aerophilic) and the gangue minerals water-avid (aerophobic). This results in separation by transfer of the valuable minerals to the air bubbles which form the froth floating on the surface of the pulp.

Flotation has a long history. The early application of flotation is relatively simple thin film flotation method, mostly using the natural hydrophobicity (or natural floatability) of minerals.

In ancient China, the natural hydrophobicity of mineral surface was used to purify mineral drugs such as cinnabar and talc. The fine mineral powder of mineral drugs was floated on the water surface to separate from the sinking gangue. In the process of washing sand gold, the feather is dipped in oil to capture the oleophilic and hydrophobic fine particles of gold and silver, called goose feather scraping gold. This method is still in use today. It is recorded in “*Heavenly Creations*” of the Ming Dynasty: in some workshop of gold and silver, people usually dropped a few clean oil in the process of recovering powder of gold and silver from waste utensils and dust, then the powder would gather together and fall to the bottom.

There was also the evidence indicated that oil and bitumen were used to collect minerals in ancient Greece and Europe. In the 18th century, it was known that solid particles stuck to gas and rose to the water surface. In the 19th century, people used to float graphite with bubbles generated by gasification (boiling pulp) or use acid react with carbonate minerals. In the late 19th century, due to the increasing demand for metals and the decreasing coarse-grained resources of lead, zinc and copper sulfide ores that can be treated by gravity separation, in Australia, the United States and some European countries, flotation has been used to separate fine ores to provide concentrate for smelting. The applied method at the initial stage is thin film flotation method or full oil flotation method. The former method is to sprinkle the ore powder on the water surface flowing in the flotation machine, and the hydrophobic minerals float on the surface and are recovered; the latter method is mixed with a certain amount of mineral oil in the pulp, and the hydrophobic and lipophilic mineral particles are stuck and float to the surface of the pulp for recovery.

At the beginning of the 20th century, the application of flotation has developed by leaps and bounds. The froth flotation technology appeared by using bubbles to increase the liquid-gas interface and improve the separation efficiency, marking the beginning research of modern flotation theory. In 1925, the collector "xanthate" was discovered, marking that flotation entered the "modern industrial application stage". Subsequently, the flotation theory, process and equipment developed rapidly and experienced a historical process from sulfide ore to non-sulfide ore, from easy-floating minerals to difficult-floating minerals, from small-scale to large-scale, from simple process to complex process.

1.2 Flotation Application

Flotation is widely used in mineral separation, known as "universal beneficiation technology".

Flotation can be used to recover low-grade and complex minerals (secondary resources) that were previously considered as no industrial value. It was initially used for sulfide minerals, and gradually developed to be used for oxidized minerals, some non-metallic minerals, coal, graphite, etc. At present, billions tons of minerals were treated by flotation in the world every year.

With the increasingly poor ore resources, the distribution of useful minerals in the ore is becoming more and more fine and miscellaneous. At the same time, the requirements and precision of the material and chemical industries for the separation of fine and ultra-fine materials are becoming higher and higher. The flotation method is becoming more and more superior to other methods and has become the most widely used and most promising beneficiation method at present.

Flotation technology has expanded to the fields of environmental protection, metallurgy, papermaking, agriculture, chemical industry, food, materials, medicine, biology, etc. With the improvement of flotation process and method, the emergence of new and high-efficiency flotation reagents and equipment, flotation method will be widely used in more industries and fields.

The effective separation particle size of conventional froth flotation is 0.01-0.3mm (0.03-0.5mm for coal). If not, the effect of froth flotation became poor at the range of other particle size. With the development of flotation technology in recent years, froth flotation method developed has a good separation application prospect for 0.5-1.0mm particles. For fine particles smaller than 0.01-0.03mm, there are a series of successful separation processes today used such as electrophoresis separation, ion flotation, etc., that using phase interface to realize separation of smaller particles such as colloidal particles, ions and molecules.

Some important points should be noted in the flotation process: the high reagent cost (compared with magnetic separation and gravity separation); strict requirements of particle size; reagents are harmful to the environment.

1.3 Flotation Machines

Flotation machine is an important equipment to realize flotation separation process (Figure 1-1). First, the pulp and the reagent react in stirring bucket. Next, the mixture is sent to the flotation machine to be stirred and aerated. Then, the useful minerals will be attached to bubbles to form mineralized bubbles. Followed, large amounts of the mineralized bubbles float to the surface of the pulp to form mineralized froth layer. At last, the froth product is scraped out with a scraper (or by overflowing) and the non-froth product is discharged from the bottom of the tank. So, the technico-economical index of flotation is closely related to the performance of flotation machine.

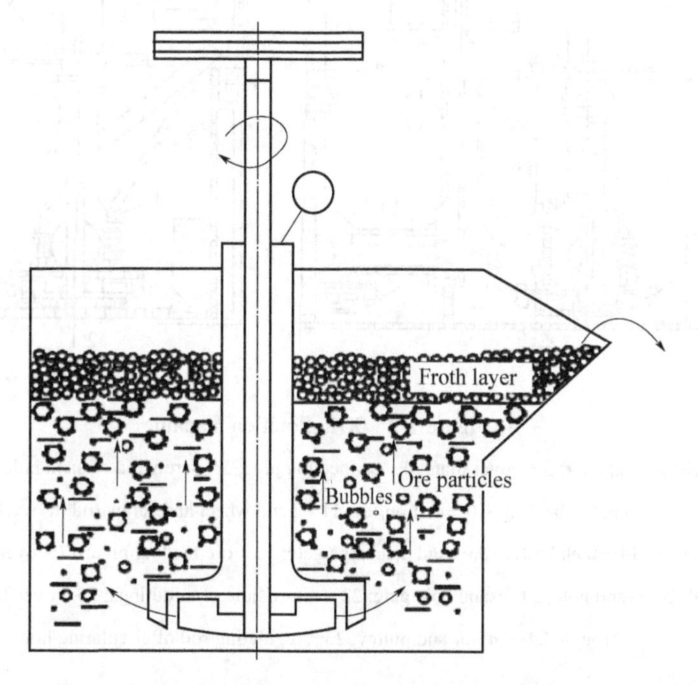

Figure 1-1 Flotation process in flotation machine

There are many kinds of flotation machines. According to different aeration and stir methods, the flotation machines can be divided into the following basic types.

(1) Mechanical agitation flotation machine. The common point of this kind of flotation machine is that the aeration and agitation of pulp are realized by mechanical mixer (rotor and stator group, the so-called structure of aeration and agitation), so it is called mechanical agitation flotation machine.

Because of the different structures of mechanical mixers such as centrifugal impeller, bar impeller, cage rotor impeller, star impeller, etc., there are also many types of mechanical agitation flotation machine such as XJK flotation machine (Figure 1-2), JJF flotation machine, SF flotation machine, CLF flotation machine, rod type flotation machine, etc.

(2) Pneumatic agitation flotation machine. The pneumatic agitation flotation machine is not only equipped with a mechanical agitator, but also equipped with the external special fan to suck air forcibly.

There are many types of pneumatic agitation flotation machine such as CHF-X14 flotation machine (Figure 1-3), AGITAIR flotation machine, BFP flotation machine, BFR flotation machine, XCF flotation machine, KYF flotation machine and YX pre-flotation machine, etc.

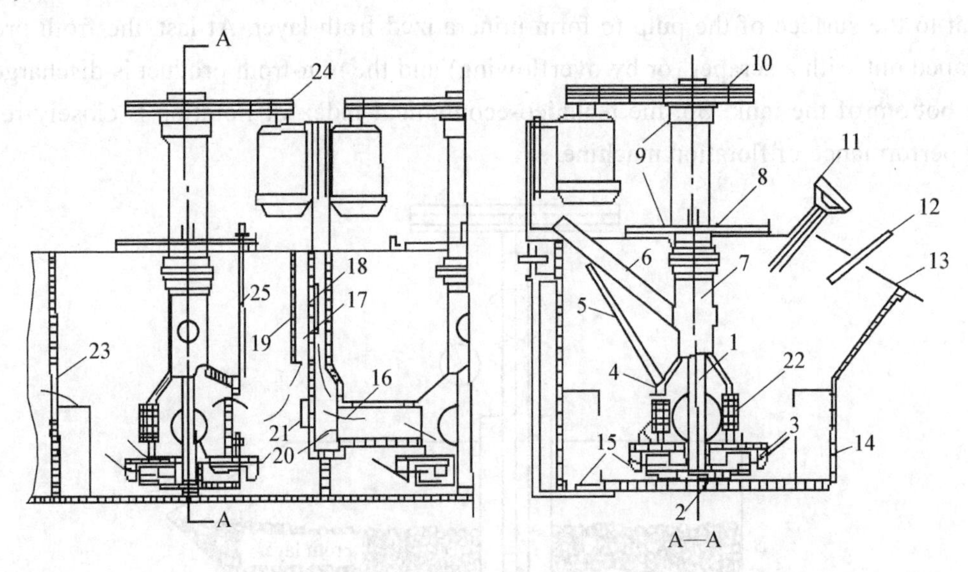

Figure 1-2 XJK flotation machine

1—principal axis; 2—impeller; 3—cover plate; 4—connecting pipe; 5—screw rod of sand hole gate; 6—air inlet pipe; 7—air pipe; 8—seat plate; 9—bearing; 10—belt pulley; 11—hand wheel and screw rod of overflow gate; 12—scraper; 13—froth overflow lip; 14—tank body; 15—sand discharge gate; 16—ore feeding pipe; 17—overflow weir; 18—overflow gate; 19—gate shell; 20—sand hole; 21—sand hole gate; 22—return hole of middlings; 23—overflow weir in front of straight trough; 24—motor and pulley; 25—regulating rod of circulating hole

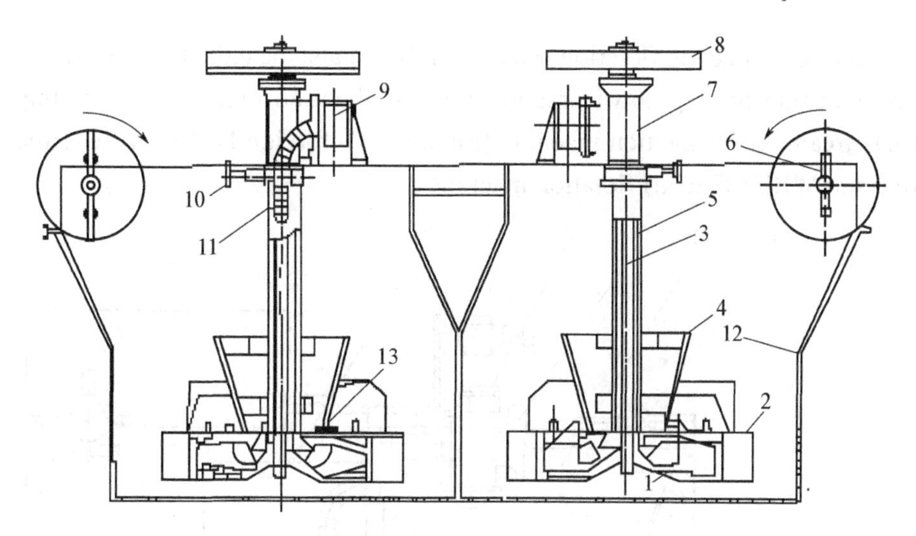

Figure 1-3 CHF-X14 flotation machine

1—impeller; 2—cover plate; 3—principal axis; 4—circular cylinder; 5—center cylinder; 6—bubble scraper; 7—bearing seat; 8—pulley; 9—main air duct; 10—control valve; 11—gas tube; 12—tank body; 13—bell

(3) Pneumatic flotation machine. Pneumatic flotation machine has no mechanical agitator and rotatable parts. The aeration of the pulp is realized by input compressed air from the external pressure fan, such as flotation column (Figure 1-4).

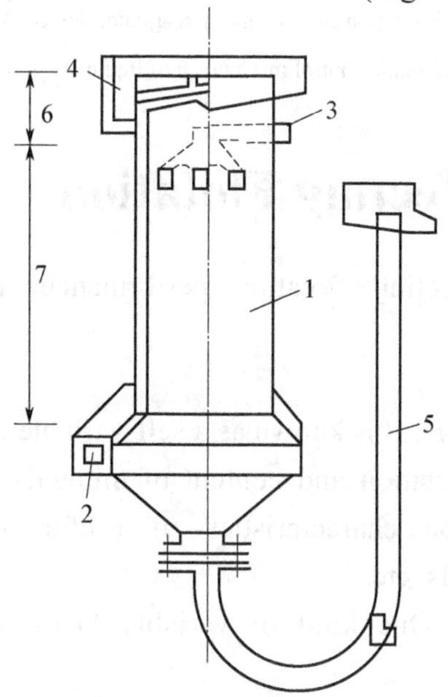

Figure 1-4 Flotation column

1—barrel; 2—compressed air input pipe; 3—ore feeding device; 4—concentrate froth flow tank; 5—tailings device; 6—concentration area; 7—collection area

(4) Air-releasing flotation machines. This is a kind of flotation machine which can release a large number of microbubbles from the solution by changing the pressure in the

flotation cell to realize the flotation separation. It can also be called variable pressure type of flotation machine or step-down pressure type of flotation machine. For example, China's XPM jet-stream whirl flotation machine (Figure 1-5), foreign DAVORA jet-stream flotation machine and WEDAG whirl flotation machine.

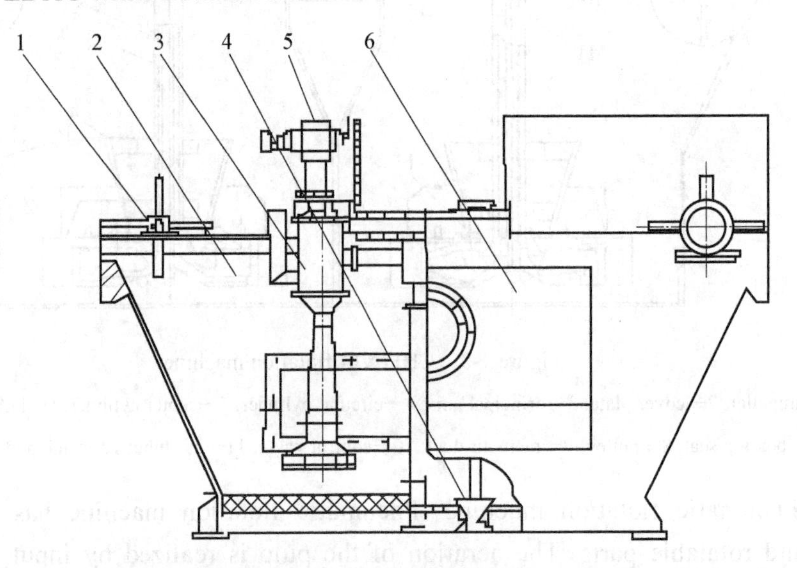

Figure 1-5 XPM jet-stream whirl flotation machine

1—scraper; 2—flotation cell; 3—gas and agitator device; 4—ore drawing device; 5—automatic control mechanism of liquid level; 6—feeding box

1.4 Factors Affecting Flotation

The main factors affecting flotation performance include process factors and technology factors.

(1) Process factors

① Non-adjustable factors: It is known as a self-variable factor.

For example, the composition and content of minerals, the degree of oxidation and argillization, the distribution characteristics of useful minerals and the symbiotic characteristics among minerals, etc.

② Adjustable factors: One kind of variable factor is to control the separation conditions.

For example, grinding fineness, pulp properties, flotation time, reagent system, flotation process, type of flotation equipment, etc.

(2) Technology factors

① The tensions of air-solid, solid-liquid and liquid-solid interfaces.

② Bubble attachment.

③ The size and shape of solids.

④ The size of bubbles.

⑤ The hydrodynamics of the system.

⑥ Chemical interaction in solution.

Those chemical factors (or phenomena) control adsorption. It is most important for the selective separation.

1.5 Flotation Chemistry

With the development of quantum chemistry, surface and colloidal chemistry, complex chemistry, organic structure theory, solid state physics and computer science, the flotation theory has been studied at the micro level and flotation chemistry has become the most important theoretical basis of flotation. Four systematic theoretical systems have been formed: flotation solution chemistry of non-sulfide ore, flotation electrochemistry of sulfide ore, flotation interface force theory of fine particle and design theory of flotation agent molecular. The physical and chemical properties of mineral surface and flotation behavior are studied from different angles with different theoretical systems.

In 1986, Professor P. Somasundaran of Columbia University published the first paper on solution chemistry of flotation. In 1988, the first monograph of "flotation solution chemistry" was published. Solution chemistry of flotation is to study the influence of mineral-solution equilibrium, flotation reagent-solution equilibrium and flotation reagent-mineral interaction equilibrium on the flotation process, so as to determine the effective components of flotation reagent for mineral flotation and the best conditions for the interaction between flotation agent and mineral.

In 1953, Salamy and Nixon pointed out that the chemical action on the surface of sulfide ore could be explained according to the electrochemical mechanism—a new direction in the field of flotation chemistry was inaugurated, named "flotation electrochemistry of sulfide ore". Since the 1970s, flotation electrochemistry has attracted the attention of scientists of mineral processing and the electrochemical mechanism of sulfide ore has been studied extensively based on electrochemical methods such as cyclic voltammetry polarization curve and AC impedance, energy band theory and molecular orbital theory. Then, the mixed potential model was put forward and the relationship between mixed potential model and electrochemical flotation was revealed. For a long time, the main research topics of flotation electrochemistry of sulfide ore are: ①the floatability of sulfide ore without collector; ②the effect of oxygen on flotation of sulfide ore; ③the action mechanism of flotation reagent and sulfide ore.

Since 1960s, with the continuous development of easily separable mineral resources, difficult, complex, poor and fine resources have become the dominant ones. Then, the interfacial force theory of fine particle flotation was built based on the classic DLVO theory and the EDLVO theory of colloidal chemistry, which mainly studied the behavior of particle agglomeration and dispersion, the interaction force between particles and bubbles or between particles.

Flotation reagent is one of the key factors of successful flotation. Research on the action mechanism of flotation reagent and development of new high-efficiency flotation reagent have always been the focus of mineral processing science and technology workers. Therefore, the design theory of flotation reagent molecular is also an important part of modern flotation chemistry.

Chapter 2
Flotation Theory

Flotation is a physico-chemical separation process that utilizes the difference in surface properties of the valuable minerals and the unwanted gangue minerals. The theory of froth flotation is complex, involving three phases (solid, water, and froth) with many subprocesses and interactions.

2.1 Theory of Surface Hydrophobicity

Froth flotation involves the aggregation of air bubbles and mineral particles in an aqueous medium with subsequent levitation of the aggregates to the surface and transfer to a froth phase as shown in Figure 2-1. Whether or not bubble attachment and aggregation

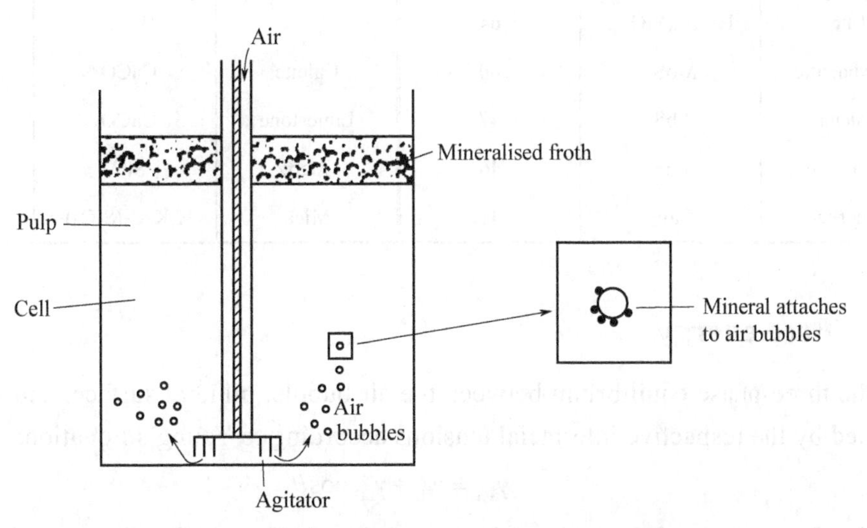

Figure 2-1 Principle of froth flotation

occur is determined by the degree to which a particle's surface is wetted by water. When a solid surface shows little affinity for water, the surface is said to be hydrophobic, and an air bubble will attach to the surface.

2.1.1 Contact angle

The hydrophobicity of mineral can be measured by the contact angle, θ, shown in Figure 2-2, which can also indicate the stability of the attachment. When the air bubble does not displace the aqueous phase, the contact angle is zero. On the other hand, complete displacement of the water represents a contact angle of 180°. Values of contact angle between these two extremes provide an indication of the degree of surface hydration, or, conversely, the hydrophobic character of the surface.

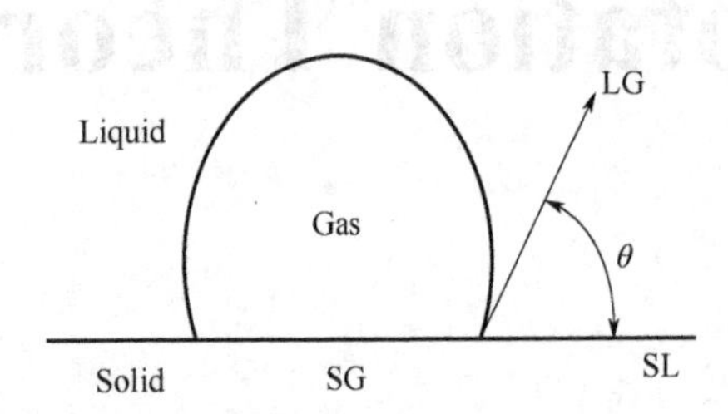

Figure 2-2 Schematic representation of the equilibrium contact between an air bubble and a solid immersed in a liquid

There are no known solids that exhibit a contact angle greater than 108° which is the value obtained with Teflon. There are a few naturally hydrophobic minerals, for example, coal, molybdenite, sulfur and talc, all of which exhibit contact angles less than 108°. Table 2-1 shows the contact angle of some minerals. Most minerals are hydrophilic and, as such, must acquire their hydrophobic character by the adsorption of surfactants, termed collectors, in order for air bubble attachment to occur.

Table 2-1 Contact angle of some minerals

Mineral	Formula	θ/(°)	Mineral	Formula	θ/(°)
Sulfur	S^0	78	Pyrite	FeS_2	30
Talc	$H_2Mg_3(SiO_3)_4$	64	Barite	$BaSO_4$	30
Molybdenite	MoS_2	60	Calcite	$CaCO_3$	20
Galena	PbS	47	Limestone	$CaCO_3$	0-10
Sphalerite	ZnS	46	Quartz	SiO_2	0-4
Fluorite	CaF_2	41	Mica	$H_2KAl_3(SiO_4)_3$	~0

2.1.2 Free energy

The three-phase equilibrium between the air bubble, mineral surface, and water can be described by the respective interfacial tensions according to Young's Equation:

$$\gamma_{SG} = \gamma_{SL} + \gamma_{LG}\cos\theta \tag{2-1}$$

The free energy change per unit area corresponding to the attachment process (the displacement of the water by the air bubble) is referred to as Dupre's Equation:

$$\Delta G = \gamma_{SG} - (\gamma_{SL} + \gamma_{LG}) \tag{2-2}$$

The free energy change can then be expressed in terms of the contact angle:

$$\Delta G = \gamma_{LG}(\cos\theta - 1) \tag{2-3}$$

And the attachment process is seen to be spontaneous for all finite contact angles. A detailed discussion of wetting and the interpretation of contact angle measurements are presented.

The free energy change for the bubble attachment process, ΔG, can also be described in terms of the work of adhesion , W_A, and the work of cohesion, W_C of water.

$$\Delta G = W_A - W_C \tag{2-4}$$

For attachment to be effected, the work of adhesion of water, W_A, must be less than the work of cohesion of water, that is

$$W_A < W_C \tag{2-5}$$

The work of adhesion is defined as the work required to remove liquid from the solid surface leaving an adsorbed water layer in equilibrium with a saturated gas phase. The work of adhesion has been shown to consist of three components:

W_I—ionization energy, arising from coulombic attractive forces at the solid surface.

W_H—hydrogen bond energy, arising from coordination forces, the dipole interaction of the solvent with the solid surface.

W_D—dispersion energy, arising from solvent inter-action with induced dipoles at the solid surface.

$$W_A = W_I + W_H + W_D \tag{2-6}$$

The work of cohesion for water is simply the energy to create new surface at the air/water interface, i.e. 2 γ_{LG} or 146 erg/ cm^2 (1erg=10^{-7}J). Of this approximately 30 percent is due to dispersion forces, and the balance is due to polar forces, principally hydrogen bonding. Since W_D less than 146 erg/ cm^2 for solids, all solids would be naturally hydrophobic if only dispersion forces were involved. The presence of ionic groups and hydrogen bonding, however, results in the wetting of most solid surfaces.

In the case of oxygen-bearing minerals, the energies of ionization and hydrogen bonding are large relative to the energy of dispersion and to the work of cohesion. In the case of silica, for example, W_D=102 erg/ cm^2 and $W_I + W_H$ = 368 erg/ cm^2 which $W_H \approx W_I$ except at high values of pH. These strong forces result from the interaction between water and silanol groups, which form when freshly-formed silica surfaces are exposed to water. The spacing of these groups is such that a number of them can form hydrogen bonds with the same water molecule. Considerable surface hydration forces are established under these conditions, and the surface is termed hydrophilic. A similar argument can be considered for ionic solids for which the ionic character of the solid accounts for the surface hydration forces.

In addition to these phenomena which determine whether or not water is displaced from the surface by an air bubble, charge separation between the solid and aqueous phases occurs, and the solid acquires a surface charge with respect to the aqueous phase. Frequently, the mobility of the charge on the solid phase is restricted and limited to surface lattice atoms, whereas the charge in the aqueous phase is mobile and distributes itself in a region adjacent to the solid surface.

2.2 Theory of Surface Charge

2.2.1 Electrical charge

Electrical charge can be generated on a solid surface by a number of mechanisms. These include: specific chemical interaction, preferential dissolution of surface ions and lattice substitution.

(1) Specific chemical interactions. Specific chemical interactions between the surface species or solute and solvent (H_2O) is termed chemisorption. These interactions include reactions with the aqueous phase which lead to the formation of different surface compounds or species. One of the most common mechanisms of charge generation, operative in many systems including oxides, silicates, and semisoluble salts, is the formation and subsequent dissociation of surface acid groups. As shown in Figure 2-3, in the case of quartz, surface silicic acid or silanol, dissociates to release hydrogen ion to solution, leaving the surface with a negatively-charged silicate group. If the hydrogen ion activity is increased, the reaction is reversed, and at high hydrogen ion activiy, the surface becomes positively charged. Hydrogen ion is said to be potential determining, then, because the surface charge and surface potential are determined by its activity in the bulk phase. At some intermediate hydrogen ion activity, the surface is uncharged, and this particular activity represents the point-of-zero-charge, pzc/PZC, of the solid. In the case of quartz the pzc occurs at pH 1.8. The pzc for any given oxide is dependent on the nature of the hydrolysis reaction for that oxide or its tendency to act as a weak acid. The specific value of the pzc will be determined by the magnitude of the dissociation constant for surface acid groups. Points-of-zero-charge which arise from this phenomenon are

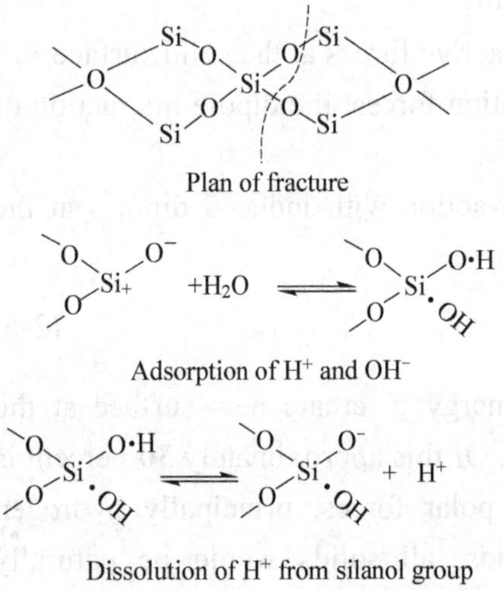

Plan of fracture

Adsorption of H^+ and OH^-

Dissolution of H^+ from silanol group

Figure 2-3 Schematic representation of surface charge development on quartz

listed for various oxide types in Table 2-2.

Table 2-2 Points-of-zero-charge arising from surface hydrolysis reactions for various types of oxides

Oxide type	pzc(pH)	Examples	pzc(pH)
M_2O	>11.5	Ag_2O	11.2
MO	8.5-12.5	MgO,	12.4
		NiO,	10.4
		CuO,	9.5
		HgO	7.3
M_2O_3	6.5-10.4	Al_2O_3,	9.1
		Fe_2O_3,	8.5
		Cr_2O_3	7.0
MO_2	0-7.5	UO_2,	6.0
		SnO_2,	4.7
		TiO_2,	4.7
		SiO_2	1.8
M_2O_5, MO_3	<0.5	WO_3	0.3

In systems involving a pure oxide in the absence of foreign polyvalent cations in solution, this mechanism is difficult to distinguish from the model in which surface charge is attributed to the adsorption of hydroxyl complexes.

Although hydrogen ion adsorption is cited as one of the common examples of specific chemical interaction, surface chemical reactions, such as hydroxyl ion adsorption, xanthate adsorption on galena, oleate adsorption on calcite, and adsorption of ions comprising the solid, would be included in this category.

In most of these systems, the surface charge density will be determined by the extent of the chemisorption reaction and the type of surface compound formed.

(2) Preferential dissolution of ions. In the absence of specific chemical interactions, solids can acquire a surface charge by preferential dissolution of surface ions. Basically, the premise is that for simple uni-univalent ionic solids, which must have equal surface distribution of cations and anions on a cleavage plane, the sign of the surface potential developed in a saturated solution is determined by the relative magnitudes of the free energies of hydration of the ions which constitute the crystal lattice. The ion with the greater negative hydration energy tends to hydrate to the greater extent and leave the surface with an excess of the opposite ion, whereby the sign of the surface charge is established. This concept is demonstrated for the silver halides as shown in Table 2-3.

The correlation between predicted and experimentally determined surface charge for uni-univalent compounds is good. However, this analysis is limited to ionic solids containing singly-charged species only. For more complex solids, i.e. those containing multivalent ions, surface binding energies must be taken into account.

Table 2-3 Sign of surface charge for silver halides predicted from the hydration free energy of gaseous ions

Salt	K_{sp}	pzc	$-\Delta G_h^-$ /(kcal/mol)	$-\Delta G_h^+$ /(kcal/mol)	Sign of surface potential	
					Predicted	Experimental
AgCl	1.7×10^{-10}	4.0	83.0	105.4	Negative	Negative
AgBr	5.0×10^{-13}	5.4	76.0	105.4	Negative	Negative
AgI	8.5×10^{-17}	5.5	66.7	105.4	Negative	Negative

Note: Free energies of gaseous ions are used is this analysis because for simple uni-univalent ionic solids with equivalent distribution of cations and anions on the surface, the lattice energy contribution to the hydration of a surface ion is equivalent for both cations and anions.(1cal=4.1868J)

(3) Lattice substitution. A third mechanism whereby a solid surface may acquire a potential is by lattice substitution. The replacement of aluminum for silicon in clays, for example, is responsible for the difference in electrical characteristics observed between the faces and edges of these minerals.

Considerable study has been given to the influence that additions of specific reagents to solids have on the properties of these solids. The additions result in small departures from stoichiometry and have a significant effect on solids' ionic conductivity diffusion and magnetic properties. This defect structure has been found to influence the electrokinetic and flotation behavior of certain minerals.

2.2.2 Electrical double layer

In the development of surface charge, whether by specific chemical interaction, preferential dissolution of surface ions, or lattice substitution, the solid surface acquires a potential with respect to the solution. The surface charge is compensated by an equal charge distribution in the aqueous phase. The charge in solution together with the charge on the solid surface is referred to as the electrical double layer. A schematic representation and potential drop across the double layer are presented in Figure 2-4. The shown are potential determining ions at the surface, a layer of adsorbed counter ions, and counter ions arranged diffusely in the solution surrounding the solid. Potential determining ions are those ions which establish the surface charge. These ions include ions of which the solid is composed, hydrogen and hydroxyl ions, collector ions that form insoluble metal collector salts with ions comprising the mineral surface, and ions capable of forming complex ions with surface species.

From the standpoint of electroneutrality, charge density in the diffuse layer, σ_d, must equal charge density on the surface, σ_s.

$$\sigma_s = -\sigma_d \tag{2-7}$$

The potential difference between the surface and the bulk solution is termed the total double layer potential, φ_0. The potential difference between a hypothetical plane representing the closest distance of approach of hydrated counter ions to the surface and the bulk solution is φ_δ, which is generally assumed to be the zeta potential.

It is not possible to measure φ_0 for non-conductive solids. However, it is possible to calculate this value once the pzc for the solid is known. That is, the dependence of φ_0 on the activity of potential determining ions is:

$$\varphi_0 = \frac{RT}{nF}\ln\frac{a_+}{a_+^0} = \ln\frac{a_-}{a_-^0} \tag{2-8}$$

Where R is the gas constant, T is absolute temperature, n is valence, F is the Faraday constant, a_+ and a_- are the activities of potential determining ions in solution and a_+^0 and a_-^0 are the activities of potential determining ions at the pzc. Using quartz as an example, the pzc of this mineral is pH 1.8. The activity of hydrogen ion at this pH is 1.58×10^{-2}, consequently, the value of φ_0 at pH 7.0 is:

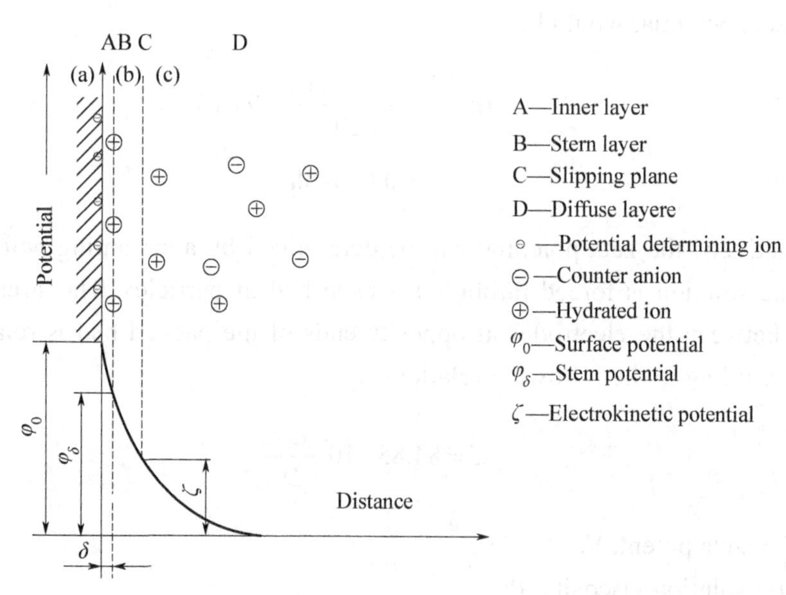

Figure 2-4 Schematic representation of the double layer and potential drop across the double layer

(a) surface charge; (b) stern layer; (c) diffuse layer of counter ions

$$\varphi_0 = \frac{1.98\text{cal}/(\text{K}\bullet\text{mol})\times298\text{K}}{1\text{mol}^{-1}\times23060\text{cal}/\text{V}}\ln\frac{1\times10^{-7}}{1.58\times10^{-2}}$$

$$\varphi_0 = 0.059(\text{PZC}-\text{pH})$$

$$\varphi_0 = 0.305\text{V}$$

Since surface potentials are not measured directly, the electro-kinetic behavior of mineral particles is characterized by zeta potential measurements. These measurements are made either by electrophoresis or by a streaming potential technique. In the case of electrophoresis, the electrophoretic mobility is measured by tracking the velocity of the particles in the absence of convective flow under a potential gradient. The zeta potential is calculated from the relationship:

$$\zeta = 4\pi \frac{\mu V}{D E} \times 9 \times 10^4 \tag{2-9}$$

Where ζ —zeta potential, V;

μ —solution viscosity, P (1P=10^{-1}Pa • s);

D—dielectric constant;

V—particle velocity, cm/s;

E—potential gradient, V/cm.

At 20℃, solution viscosity would be 0.01 poise and dielectric constant would be 78. Assuming $V=7.3\times10^{-3}$cm/s and that the applied voltage is 200 volts across a cell length of 10cm, the zeta potential would be:

$$\zeta = 4\pi \frac{7.3\times10^{-3}\times0.01}{78\times20}\times9\times10^4$$

$$\zeta = 0.053\text{volt}$$

Alternatively, the zeta potential can be determined by a streaming potential technique in which the solution is forced through a packed bed of particles. The streaming potential developed between the electrodes at opposite ends of the packed bed is related to the zeta potential according to the following relationship:

$$\zeta = 84.85\times10^7 \frac{\mu\sigma E_s}{D\Lambda_p} \tag{2-10}$$

Where ζ —zeta potent, V;

μ—solution viscosity, P;

D—dielectric constant;

σ—conductivity, S/cm;

Λ_p —pressure drop, cm Hg (1cm Hg=1333.22Pa);

E_s —streaming potential, V.

The majority of solid possess an electrical charge on their surfaces in aqueous medium with the consequence that extensive hydration of their surfaces occurs. As a result, collectors must be added to render the surfaces sufficiently hydrophobic for air bubble attachment to occur.

2.3 Theory of Mineral Adsorption

2.3.1 The kind of adsorption

Adsorption is a phenomenon of some substance of the liquid (or gas) concentrates (or disperses) at the phase interface.

The adsorption of collectors and regulators on the solid-liquid interface directly affects the physical and chemical properties of the mineral surface, which can regulate the floatability of minerals.

Classification according to adsorption characteristics:

(1) Molecular adsorption: Adsorption of molecules dissolved in solution on the surface. For example, the adsorption of frother molecules of terpineol oil or alcohol at the liquid-gas interface; adsorption of the weak electrolyte molecules and neutral oil molecules on the mineral surface, etc.

(2) Ion adsorption: Adsorption of certain ions in the solution on mineral surface. For example, the adsorption of xanthate ion on the surface of sulfide ore, and the adsorption of Ca^{2+} on the surface of quartz.

(3) Exchange adsorption: Some ions in the solution are adsorbed on the mineral surface by exchanging with another ion of the mineral surface. For example, Cu^{2+} in the solution can exchange with Zn^{2+} of the surface lattice of sphalerite to activate sphalerite and improve its floatability.

(4) Inlayer adsorption of electrical double layer (localized adsorption): The lattice ions, lattice isomorphic ions or localized ions in the solution are adsorbed on the mineral surface and this results in the change of potential (value or symbol). For example, the adsorption of Ba^{2+} and SO_4^{2-} ions on barite surface and the adsorption of H^+ and OH^- on quartz surface.

(5) Out-layer adsorption of electrical double layer: Adsorption of molecules or ions of solute in solution on the out-layer of electric double layer. This adsorption depends entirely on the action of electrostatic gravity. It can only change the size of the zeta potential, but not change the sign of the potential. Such adsorption can be produced by any ion with the opposite sign of the surface charge of the mineral.

(6) Hemimicelles adsorption: When the concentration of the collector with long hydrocarbon chain in the solution is high, the non-polar groups of the collector adsorbed on the mineral surface associate with each other by the action of van der Waals force.

(7) Characteristic adsorption: The adsorption phenomenon caused by the affinity of mineral surface for some component in solution.

According to the nature of adsorption, it can be divided into physical adsorption and chemical adsorption.

Physisorption: All adsorption caused by molecular bond force (van der Waals force).

The characteristics are:

(1) Small thermal effect (generally about 21kJ/mol)

(2) Reversibility (The adsorbate is easily desorbed from the surface)

(3) No selectivity

(4) The adsorption speed is fast

(5) Multilayer adsorption (multilayer molecules or ions)

For example, molecular adsorption, out-layer adsorption of electrical double layer and hemimicelles adsorption.

Chemisorption: Any adsorption caused by chemical bond forces.

The characteristics are:

(1) Large thermal effect(generally between 20-200 kJ/mol)

(2) Irreversibility (the adsorption is firm, not easy to desorb)

(3) Monolayer adsorption

(4) Strong selectivity

(5) The adsorption speed is slow

For example, exchange adsorption, localized adsorption, etc.

Chemical adsorption is different from chemical reaction. Chemical adsorption cannot form a new "phase", and the components of adsorption products are different from the chemical reaction products.

2.3.2 CMC's implication to flotation

Counter ions are those ions which have no special affinity for the surface and are adsorbed by electrostatic attraction. Examples are chloride ion and collector anions at low concentration adsorbed on a positively charged surface.

Certain ions possess special affinity for the surface but are not chemisorbed, and these are termed specifically adsorbed ions. Examples include polyvalent metal ions in a pH region in which the metal ions hydrolyze to hydroxyl complexes and also longer-chained homologues of collectors at moderate to high concentration. At certain concentrations the hydrocarbon chains of the surfactant ions and/or molecules associate and micellization occurs. See Figure 2-5, the concentration at which this phenomenon occurs is the critical micelle concentration and is referred to as the CMC. When this phenomenon occurs at the mineral surface, these aggregates are termed hemimicelles.

Depending on concentration and hydrocarbon chain length, collector ions may be present on the surface as individual ions or as aggregates of ions, termed hemi-micelles. These phenomena are demonstrated well by the data of Figure 2-6, at relatively low concentrations of dodecyl sulfonate, i.e. less than about 5×10^{-5} mole. Individual sulfonate ions adsorb as counter ions at the solid-liquid interface by electrostatic attraction. This phenomenon occurs in Region Ⅰ and is shown schematically in Figure 2-7(a). As the concentration of dodecyl sulfonate is increased, the adsorption density of surfactant becomes sufficiently high that interaction between hydrocarbon chains of the sulfonate ions occurs through van der Waals forces. This phenomenon results in hemimicelle formation and occurs at concentrations designated as Regions Ⅱ and Ⅲ. The solid-liquid interface under these conditions is depicted schematically in Figure 2-7(b). The break between Region Ⅱ and Region Ⅲ occurs at about 1/10 monolayer coverage.

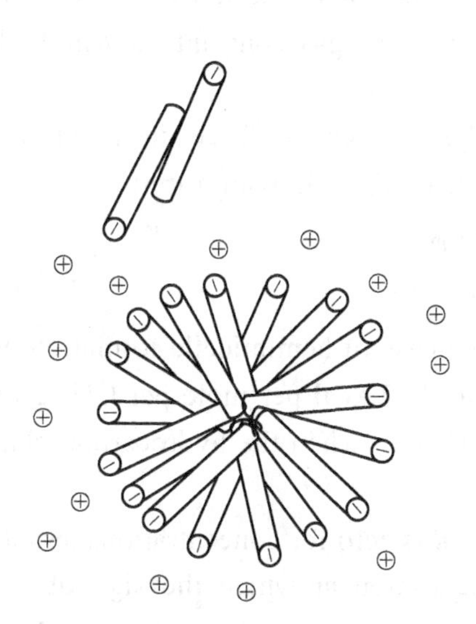

Figure 2-5 Schematic representation of an anionic micelle

Figure 2-6 Adsorption density and zeta potential of alumina as a function of dodecyl sulfonate concentration at pH 7.2 and 2×10^{-3} mol/L ionic strength

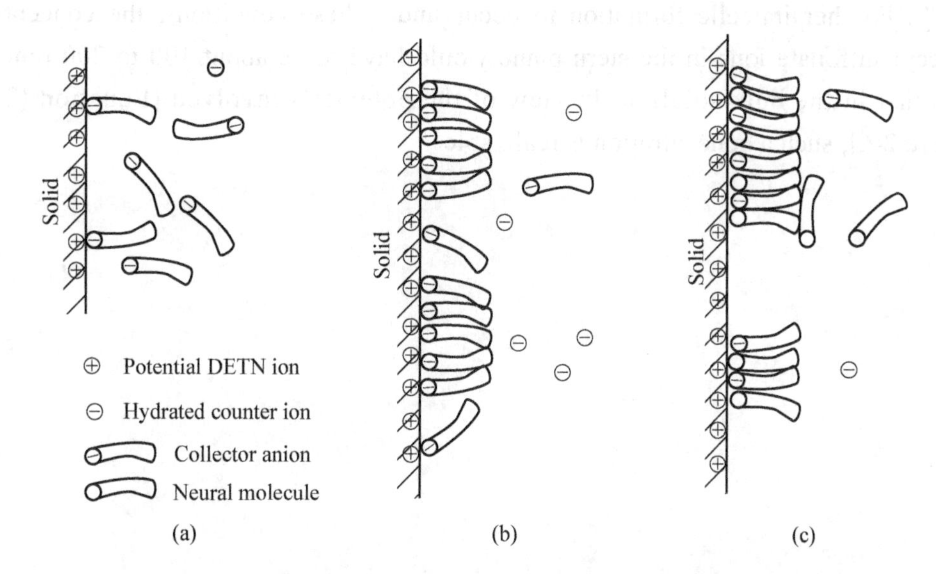

Figure 2-7 Adsorption of collector anion on electric double layer (DETN=determining)

(a) low concentration; (b) high concentration, forms hemimicelles; (c) adsorption of collector anions and molecules

The adsorption density of sulfonate ions adsorbed in Region Ⅰ may be calculated by the following relationship:

$$\Gamma_i = 2r_iC_i\exp\left(-\frac{nF\zeta}{RT}\right) \tag{2-11}$$

Where Γ_i is the adsorption density of species i (mol/cm^2); r is the radius of the adsorbed counter ion (cm); C is the concentration in the bulk solution (mol/cm^3); n is valence of the ion ; ζ is the zeta potential (volt); R is the gas constant [cal/(mol • K), 1cal=4.1868J]; and F is the Faraday constant (cal/V).

When species are adsorbed specifically at the surface, a force in addition to electrostatic force is involved, and Equation (2-12) takes on the following form:

$$\Gamma_i = 2r_iC_i\exp\left(-\frac{nF\zeta + \phi}{RT}\right) \tag{2-12}$$

Where ϕ is the specific adsorption potential. In the case of hemimicelle formation, van der Waals forces assume importance, and ϕ is equal to 0.62 kcal per mole per CH_2 group. This is the free energy decrease that the system experiences when the hydrocarbon chains are removed from water.

With reference to the data presented in Figure 2-6, ϕ is zero for concentrations less than about 5×10^{-5} molar sulfonate. At the pzc, the concentration at which the sign of ζ is reversed and in this case is about 3×10^{-4} mole,1/10 monolayer coverage is present. Under these conditions ϕ will equal 0.62 kcal/mol CH_2 group.

Adsorption data in Figure 2-6 show that hemimicelles form at the alumina surface when the bulk equilibrium concentration of dodecyl sulfonate concentration is approximately 5×10^{-5} mol/L. For hemimicelle formation to occur under these conditions, the concentration of dodecyl sulfonate ions in the stern plane would have to be about 100 to 200 times greater than that in the bulk solution. In view of the potentials involved [Equation (2-10) and Figure 2-6], such a concentration is realizable.

Chapter 3

Chemistry of Flotation Reagent

All kinds of flotation reagents, whether they are organic collectors, frothers, depressants, flocculants, or various regulators dominated by inorganic substances, play a role in pulp solution and mineral/solution interface. Their existing state in solution and their basic chemical behavior have important effects on the flotation process. The solution chemistry of flotation reagent is the basic knowledge of studying the function of flotation reagent and controlling the process of using flotation reagent, which is of great significance.

3.1 The Equilibration of Flotation Reagent in Solution

3.1.1 Acid-base theories

At the end of the 19th century, according to the ionization theory of electrolyte, acid was defined as a substance that ionized and released hydrogen ions in water, while base was a substance that ionized and released hydroxyl ions. This is the classic concept of acid and base.

In 1923, Bronsted and Lowry proposed a broad concept of acid-base. Acid is defined as A, which can give hydrogen ions (protons), and base B is a proton with a tendency to bind protons.

That is:

$$\text{Bronsted acid A} \rightleftharpoons \text{Bronsted base} + H^+ \tag{3-1}$$

Such as:

$$NH_4^+ \rightleftharpoons NH_3 + H^+$$

$$H_2C_2O_2 \rightleftharpoons HC_2O_2^- + H^+$$

There are various types of traditional acid-base reactions:

e.g:	Traditional name:
$CH_3COOH + NH_3 \rightleftharpoons CH_3COO^- + NH_4^+$	Salt formation
$CH_3COOH + H_2O \rightleftharpoons CH_3COO^- + H_3O^+$	Acid dissociation
$NH_3 + H_2O \rightleftharpoons NH_4^+ + OH^-$	Alkali dissociation
$H_3O^+ + OH^- \rightleftharpoons H_2O + H_2O$	Neutralization reaction
$NH_4^+ + H_2O \rightleftharpoons NH_3 + H_3O^+$	Weak alkali salt hydrolysis
$CN^- + H_2O \rightleftharpoons HCN + OH^-$	Weak acid salt hydrolysis
$H_2O + H_2O \rightleftharpoons H_3O^+ + OH^-$	Water dissociation

It can be seen from the above reactions that although there are various types of traditional acid-base reactions, their basic feature is that the reactions are carried out through proton transfer, so these reactions are collectively referred to as proton transfer reactions. In fact, the substances on both sides of the above reaction are the simplest flotation reagent, which means that the flotation reagent will also have the above reactions in the solution.

Lewis put forward a more generalized acid-base theory on the basis of chemical bond electron theory in 1923. Acid is defined as the acceptor of electron pair and base as the donor of electron pair. Lewis acid base contains not only all Bronsted acid-base, but also contains many acid-base of non-conforming Bronsted definition, but it does have acid-base substances, because some substances have no proton to transfer.

Such as:

$$Al^{3+} + 6H_2O \rightleftharpoons Al(H_2O)_6^{3+}$$

In this way, according to Lewis acid-base theory, most flotation reagents, except polar oil, can be regarded as acid-base. For example, inorganic cations (activators) such as Cu^{2+}, Pb^{2+}, Ca^{2+}, Mg^{2+} are acids, while some collectors and regulator anions such as CN^-, $RCOO^-$, $ROC(=S)-S^-$ are bases. Therefore, solution equilibrium calculation of flotation reagent belongs to acid-base balance calculation.

3.1.2 Balance equations for solution equilibria

The solution equilibrium equation is the basic relational equation for the calculation of the solution equilibrium of the flotation reagent, and there are mainly the following three types. Take the flotation regulator Na_2CO_3 as an example to illustrate various equilibrium relationships.

(1) Proton balance equation（PBE）

$$Na_2CO_3 + 2H_2O \rightleftharpoons H_2CO_3 + 2NaOH \quad (3\text{-}2)$$

$$H_2CO_3 \rightleftharpoons H^+ + HCO_3^- \qquad K_{a1} = \frac{[H^+][HCO_3^-]}{[H_2CO_3]} \tag{3-3}$$

$$HCO_3^- \rightleftharpoons H^+ + CO_3^{2-} \qquad K_{a2} = \frac{[H^+][CO_3^{2-}]}{[HCO_3^-]} \tag{3-4}$$

$$H_2O \rightleftharpoons H^+ + OH^- \qquad K_w = [H^+][OH^-] \tag{3-5}$$

In the formula, K_{a1}, K_{a2} and K_w are the first and second dissociation constants of carbonic acid and the ion product of water, respectively.

(2) Mass balance equation(MBE). MBE expresses the fact that the sum of the concentrations of various forms produced by an added substance is equal to the analytical concentration of the substance. MBE for Na_2CO_3 is:

$$C_{Na_2CO_3} = [H_2CO_3] + [HCO_3^-] + [CO_3^{2-}] \tag{3-6}$$

$$C_{Na_2CO_3} = [Na^+] \tag{3-7}$$

(3) Charge balance equation(CBE). Any electrolyte solution is electrically neutral, and the total concentration of positively charged components must be equal to that of negatively charged components. CBE of Na_2CO_3 is:

$$[H^+] + [Na^+] = [HCO_3^-] + 2[CO_3^{2-}] + [OH^-] \tag{3-8}$$

If MBE and CBE are combined, the algebraic operation can be simplified. From Equations (3-7) and (3-8):

$$[H_2CO_3] + [H^+] = [OH^-] + [CO_3^{2-}] \tag{3-9}$$

The above formula is called the proton balance equation of HCO_3^-(PBE).

The proton balance equation shows that the number of protons obtained by some initial substance (such as HCO_3^- and H_2O) is exactly equal to the number of protons lost by the initial substance on the other side.

Water can obtain protons to generate $H_3O^+ \approx H^+$, and lose protons to generate OH^-. HCO_3^- obtains protons to generate H_2CO_3 and loses protons to generate CO_3^{2-}. Therefore, the PBE is shown in formula (3-9).

Similarly, the PBE of H_2CO_3 and H_2O is:

$$[H^+] = [HCO_3^-] + [CO_3^{2-}] + [OH^-] \tag{3-10}$$

PBE of CO_3^{2-} and H_2O is:

$$[H^+] + [HCO_3^-] + [H_2CO_3] = [OH^-] \tag{3-11}$$

3.1.3 pH value of solution of flotation reagent

Due to the acidity and alkalinity of the flotation reagent itself, it has hydrolysis or dissociation reaction in the aqueous solution, which changes the pH value of the medium, and then affects the effect of the reagent on the floating minerals and the interaction between the reagents. It is of great significance for laboratory research and the regulation and control of pulp pH regulation and reagent interaction in the production process of flotation plant to calculate the change of medium pH value caused by a certain concentration of reagent.

(1) pH value of strong acid and strong base flotation reagent solution. There are relatively few flotation reagents belonging to the strong acid and strong base type. Strong acid such as pH regulator: HCl, H_2SO_4; collector: hydrocarbon sulfonic acid RSO_3H and hydrocarbon sulfonic acid $ROSO_3H$. Strong alkaline such as pH regulator NaOH, collector quaternary amine $RN(CH_3)_3^+$. The pH value of this type of flotation reagent solution, when the concentration is relatively dilute, can be considered as:

$$C_A = [H^+] \quad \text{or} \quad C_B = [OH^-] \tag{3-12}$$

That is, the hydrogen ion concentration in the strong acid dilute solution is equal to the analytical concentration of the acid; the hydroxide ion concentration in the strong base dilute solution is equal to the analytical concentration of the base.

(2) pH value of monobasic weak acid and monobasic weak base flotation reagent solution. Let the monobasic weak acid be HA, and write various equilibrium relationships.

Dissociation equilibrium:

$$HA \rightleftharpoons H^+ + A^- \qquad K_a = \frac{[H^+][A^-]}{[HA]} \tag{3-13}$$

Mass balance:

$$C_A = [A^-] + [HA] \tag{3-14}$$

Substitute Equation (3-13) into Equation (3-14):

$$[A^-] + [HA] = [A^-] + \frac{[H^+][A^-]}{K_a} = [H^+] + \frac{[H^+]^2}{K_a}$$

$$[H^+]^2 + K_a[H^+] - K_a C_A = 0$$

$$[H^+] = -\frac{K_a}{2} + \sqrt{\frac{K_a^2}{4} + K_a C_A} \tag{3-15}$$

For monobasic weak base B, its dissociation equilibrium:

$$BH^+ \rightleftharpoons B + H^+ \tag{3-16}$$

$$K_a = \frac{[H^+][B]}{[BH^+]} = \frac{[H^+][OH^-][B]}{[BH^+][OH^-]} = \frac{K_W}{K_b} \tag{3-17a}$$

Where $K_b = \frac{[BH^+][OH^-]}{[B]}$ is the equilibrium constant of the following reactions:

$$B+H_2O \rightleftharpoons BH^+ +OH^- \tag{3-17b}$$

K_b is the base dissociation constant.

By mass balance:

$$C_B = [B]+[BH^+] = [B]\left(1+\frac{[BH^+]}{[B]}\right) \tag{3-18}$$

From equation (3-17b), $[B]=\frac{[BH^+][OH^-]}{K_b} \approx \frac{[OH^-]^2}{K_b}$

$$\frac{[BH^+]}{[B]}=\frac{K_b}{[OH^-]}$$

So

$$C_B=\frac{[OH^-]^2}{K_b}\left(1+\frac{K_b}{[OH^-]}\right) \tag{3-19a}$$

$$[OH^-]^2+K_b[OH^-]-K_bC_B=0 \tag{3-19b}$$

$$[OH^-]=-\frac{K_b}{2}+\sqrt{\frac{K_b^2}{4}+K_bC_B} \tag{3-19c}$$

or

$$[OH^-]=-\frac{K_w}{2K_a}+\sqrt{\frac{K_w^2}{4K_a^2}+\frac{C_BK_w}{K_a}} \tag{3-19d}$$

The relationship between pH value and reagent concentration of common monobasic weak acid and monobasic weak base flotation reagent solutions calculated from Equations (3-15) and (3-19) is shown in Table 3-1.

Table 3-1 pH value of common monobasic weak acid and weak base flotation reagent solutions

Flotation reagent	Dissociation constant K_a	pH						
		10^{-1}mol/L	10^{-2}mol/L	10^{-3}mol/L	5×10^{-4}mol/L	10^{-4}mol/L	5×10^{-5}mol/L	10^{-5}mol/L
Xanthic acid	10^{-5}	3.0	3.5	4.0	4.2	4.6	4.7	5.2
Oleic acid	10^{-5}	3.5	4.0	4.5	4.7	5.0		5.6
HCN	$10^{-9.21}$	5.1	5.6	6.1		6.6		
Dodecylamine	$10^{-10.63}$	11.8	11.3	10.7		9.9		8.3
Thiophosphate	2.3×10^{-5}							

It can be seen that in the range of flotation concentration (10^{-5}-10^{-8}mol/L), xanthic acid, oleic acid and other solutions are acidic, while alkyl amine solution is alkaline.

(3) Monobasic strong base weak acid salt solution. Assuming that the monobasic weak acid salt is BA, PBE is:

$$\left[H^+\right]+[HA] \rightleftharpoons \left[OH^-\right] \tag{3-20}$$

$$\left[H^+\right]+\frac{\left[H^+\right]\left[A^-\right]}{K_a}=\frac{K_w}{\left[H^+\right]}$$

$$\left[H^+\right]^2\left(1+\frac{\left[A^-\right]}{K_a}\right)=K_w$$

then

$$\left[H^+\right]=\sqrt{\frac{K_a K_w}{K_a+\left[A^-\right]}} \tag{3-21a}$$

For strong base weak acid salt $\left[A^-\right]\approx C_{BA}$

So

$$\left[H^+\right]=\sqrt{\frac{K_a K_w}{K_a+C_{BA}}} \tag{3-21b}$$

Table 3-2 shows the pH values of common strong base weak acid salt flotation reagent solutions, indicating that these flotation reagent solutions are alkaline in varying degrees. The research shows that the effective pH range of xanthate and sodium oleate for the flotation of many minerals is 7-9, and the effective pH range of sodium hydroxamate for the flotation of many minerals is 8-9. When these reagents are in the flotation concentration range (10^{-5}-10^{-3}mol/L), the pH value in the solution is basically in their effective flotation pH range. Therefore, if other factors do not affect the pH value, it is generally unnecessary to add another pH regulator when flotation minerals with these reagents.

Table 3-2 pH value of common monobasic strong base weak acid flotation reagent solutions

Flotation reagent	Dissociation constant K_a	pH					
		10^{-1}mol/L	10^{-2}mol/L	10^{-3}mol/L	10^{-4}mol/L	10^{-5}mol/L	10^{-6}mol/L
Ethyl xanthate	10^{-5}	9.0	8.5	8.0	7.5	7.2	7.0
Ethyl dithiophosphate	2.3×10^{-5}	8.8	8.3	7.8	7.4	7.1	7.0
Sodium oleate	10^{-6}	9.5	9.0	8.5	8.0	7.5	7.2
Sodium octyl hydroxamate	10^{-9}	11.0	10.5	10.0	9.5	9.0	8.5
Sodium cyanide	$10^{-9.21}$	11.1	10.6	10.1	9.6	9.1	8.6

(4) Polybasic weak acid flotation reagent. The general formula of the polybasic weak acid flotation reagent is H_nA, and its proton transfer balance is carried out step by step:

$$A+H^+ \rightleftharpoons HA \qquad K_1^H=\frac{[HA]}{[A]\left[H^+\right]}=\beta_1^H$$

$$HA+H^+ \rightleftharpoons H_2A \qquad K_2^H=\frac{[H_2A]}{\left[H^+\right][HA]}$$

$$\beta_2^{\mathrm{H}} = \frac{[\mathrm{H_2A}]}{[\mathrm{H^+}]^2[\mathrm{A}]} = K_1^{\mathrm{H}} K_2^{\mathrm{H}}$$

$$\vdots$$

$$\mathrm{H}_{n-1}\mathrm{A} + \mathrm{H}^+ \rightleftharpoons \mathrm{H}_n\mathrm{A} \qquad K_n^{\mathrm{H}} = \frac{[\mathrm{H}_n\mathrm{A}]}{[\mathrm{H}_{n-1}\mathrm{A}][\mathrm{H}^+]}$$

$$\beta_n^{\mathrm{H}} = \frac{[\mathrm{H}_n\mathrm{A}]}{[\mathrm{H}^+]^n[\mathrm{A}]} = \prod_{l=1}^{n} K_i^{\mathrm{H}} \tag{3-22}$$

In the formula, K_1^{H}, K_2^{H}, ⋯, K_n^{H} are called stepwise proton constants, and their relationship with the dissociation constant is $K_n^{\mathrm{H}} = \frac{1}{K_{\mathrm{a1}}}$, ⋯, $K_2^{\mathrm{H}} = K_{\mathrm{a},n-1}^{-1}$, $K_1^{\mathrm{H}} = K_{\mathrm{a},n}^{-1}$.

β_1^{H}, β_2^{H}, ⋯, β_n^{H} are called accumulation plus proton constants.

Let [A], $[\mathrm{A}]'$ be the total concentration of free A and A', respectively,

$$\begin{aligned} C_{\mathrm{H}_n\mathrm{A}} &= [\mathrm{A}]' = [\mathrm{A}] + [\mathrm{HA}] + \cdots + [\mathrm{H}_n\,\mathrm{A}] \\ &= [\mathrm{A}] + \beta_1^{\mathrm{H}}[\mathrm{A}][\mathrm{H}^+] + \cdots + \beta_n^{\mathrm{H}}[\mathrm{A}][\mathrm{H}^+]^n \end{aligned} \tag{3-23}$$

If the solution contains only one multicomponent weak acid, its PBE is:

$$[\mathrm{H}^+] = [\mathrm{H}_{n-1}\mathrm{A}] + 2[\mathrm{H}_{n-2}\mathrm{A}] + \cdots (n-1)[\mathrm{HA}] + n[\mathrm{A}] + [\mathrm{OH}^-] \tag{3-24a}$$

Since the main components are $\mathrm{H}_n\mathrm{A}$ and $\mathrm{H}_{n-1}\mathrm{A}$, and $[\mathrm{OH}^-]$ is also negligible for polybasic weak acid solutions, the simplified PBE is:

$$[\mathrm{H}^+] \approx [\mathrm{H}_{n-1}\mathrm{A}] = \beta_{n-1}^{\mathrm{H}}[\mathrm{A}][\mathrm{H}^+]^{n-1} \tag{3-24b}$$

Then

$$C_{\mathrm{H}_n\mathrm{A}} \approx [\mathrm{H}_{n-1}\mathrm{A}] + [\mathrm{H}_n\mathrm{A}] = \beta_{n-1}^{\mathrm{H}}[\mathrm{A}][\mathrm{H}^+]^{n-1} + \beta_n^{\mathrm{H}}[\mathrm{A}][\mathrm{H}^+]^n \tag{3-25}$$

$$\frac{C_{\mathrm{H}_n\mathrm{A}}}{[\mathrm{H}^+]} = \frac{\beta_{n-1}^{\mathrm{H}}[\mathrm{A}][\mathrm{H}^+]^{n-1} + \beta_n^{\mathrm{H}}[\mathrm{A}][\mathrm{H}^+]^n}{\beta_{n-1}^{\mathrm{H}}[\mathrm{A}][\mathrm{H}^+]^{n-1}} = 1 + K_n^{\mathrm{H}}[\mathrm{H}^+]$$

$$K_n^{\mathrm{H}}[\mathrm{H}^+]^2 + [\mathrm{H}^+] - C_{\mathrm{H}_n\mathrm{A}} = 0 \tag{3-26}$$

Calculate the pH value of some polyacid flotation reagent solutions according to formula (3-26), as shown in Table 3-3. It can be seen that the pH value of common polyacid flotation reagent solution is relatively strong acidic. When this kind of collector is used to float minerals under neutral or alkaline conditions, or this kind of regulator is used to inhibit and disperse gangue, a certain amount of alkali needs to be added. Therefore, using this kind of flotation reagent, the adjustment and control of pulp pH value is more complex.

Table 3-3 pH value of polybasic weak acid flotation reagent solutions

Molecular formula of flotation agent	K_n^{H}	pH				
		10^{-1}mol/L	10^{-2}mol/L	10^{-3}mol/L	10^{-4}mol/L	10^{-5}mol/L
$CH_3C_6H_4As(OH)_2$		2.36	2.88	3.45	4.14	5.00
H_3PO_4	$10^{2.15}$	1.6	2.2	3.0	3.91	4.3
C_6H_8O(citric acid)	$10^{3.13}$	2.1	2.6	3.2	4.0	5.0
$C_4O_6H_6$(tartaric acid)	$10^{3.93}$	2.5	3.0	3.5	4.2	5.0
$H_2C_2O_4$(oxalate)	$10^{1.26}$	1.3	2.1	3.0	4.0	5.0

(5) pH value of polybasic weak acid and strong base salt flotation reagent solution. Let the polybasic weak acid and strong base salt be B_nA, and the PBE of A is:

$$\left[\mathrm{H}^+\right]+[\mathrm{HA}]+2[\mathrm{H_2A}]+\cdots+(n-1)[\mathrm{H}_{n-1}\mathrm{A}]=\left[\mathrm{OH}^-\right] \tag{3-27}$$

Since the main components are [A], [HA], and [H^+] is also negligible, the simplified PBE is:

$$[\mathrm{HA}]=\left[\mathrm{OH}^-\right] \tag{3-28}$$

also $$C_{\mathrm{B}_n\mathrm{A}}=[\mathrm{A}]+[\mathrm{HA}]=\frac{[\mathrm{HA}]}{K_1^{\mathrm{H}}\left[\mathrm{H}^+\right]}+\left[\mathrm{OH}^-\right]=\frac{\left[\mathrm{OH}^-\right]}{K_1^{\mathrm{H}}\left[\mathrm{H}^+\right]}+\left[\mathrm{OH}^-\right]$$

$$\left[\mathrm{OH}^-\right]^2+K_1^{\mathrm{H}}K_{\mathrm{w}}\left[\mathrm{OH}^-\right]-K_1^{\mathrm{H}}K_{\mathrm{w}}C_{\mathrm{B}_n\mathrm{A}}=0 \tag{3-29}$$

or $$\left[\mathrm{H}^+\right]^2-\frac{K_{\mathrm{w}}}{C_{\mathrm{B}_n\mathrm{A}}}\left[\mathrm{H}^+\right]-\frac{K_{\mathrm{w}}}{K_1^{\mathrm{H}}C_{\mathrm{B}_n\mathrm{A}}}=0$$

then $$\left[\mathrm{OH}^-\right]=-\frac{K_1^{\mathrm{H}}K_{\mathrm{w}}}{2}+\sqrt{\frac{\left(K_1^{\mathrm{H}}K_{\mathrm{w}}\right)^2}{4}+C_{\mathrm{B}_n\mathrm{A}}} \tag{3-30a}$$

$$\left[\mathrm{H}^+\right]=\frac{K_{\mathrm{w}}}{2C_{\mathrm{B}_n\mathrm{A}}}+\sqrt{\frac{K_{\mathrm{w}}^2}{4C_{\mathrm{B}_n\mathrm{A}}^2}+\frac{K_{\mathrm{w}}}{K_1^{\mathrm{H}}C_{\mathrm{B}_n\mathrm{A}}}} \tag{3-30b}$$

The pH value of the polybasic weak acid and strong base salt solution can be calculated from the above formula, and some results are shown in Table 3-4. These reagents are often used as depressant, dispersants or activators in flotation.

If the flotation is carried out under alkaline conditions, these reagents can be used both as modifiers and as pH control. For example, the pH required for the sulfide flotation of copper oxide and lead ore is generally 8-10. Using Na_2S as sulfurization reagent can play an activation role without adding another pH regulator.

Another example is that sodium carbonate is often used as a pH adjuster, dispersant, etc., but when the flotation needs to be carried out under strong alkaline conditions, such as the high alkaline process of Pb-Zn flotation separation, the pH must be as high as 12, then it cannot use Na_2CO_3 as pH adjuster. It can be seen from Table 3-4 that the pH value of Na_2CO_3 solution cannot reach 12 even at very high concentration.

Table 3-4 pH value of common polybasic weak acid and strong base flotation reagents

Molecular formula of flotation agent	K_1^H	pH					
		10^{-1}mol/L	10^{-2}mol/L	10^{-3}mol/L	10^{-4}mol/L	10^{-5}mol/L	10^{-6}mol/L
Na_3PO_4	$10^{12.35}$	12.6	44.9	11.0	10.0	9.0	8.0
Na_2CO_3	$10^{9.57}$	11.3	10.8	10.2	9.7	8.9	
Na_2S	$10^{13.8}$	12.9	12.0	11.0	10.0	9.0	8.0
$Na_2C_2O_4$	$10^{4.28}$	8.6	8.1	7.6	7.1		
Na_2SiO_3	$10^{12.06}$	12.4	11.8	11.0	10.0	9.0	8.0

3.1.4 Dissociation equilibrium of monobasic weak acid(base)

(1) Anionic flotation reagent. Anionic flotation reagents include monobasic weak acids and their salts. If NaA is a weak acid and strong base salt, hydrolysis reaction occurs first and then dissociation:

$$NaA + H_2O \rightleftharpoons HA + NaOH \tag{3-31}$$

$$HA \rightleftharpoons H^+ + A^- \qquad K_a = \frac{[H^+][A^-]}{[HA]} \tag{3-32}$$

Take logarithm:

$$pH - pK_a = \lg\frac{[A^-]}{[HA]} \tag{3-33}$$

The flotation significance of the above formula is to determine the conditions under which the flotation reagent has an effective electrostatic effect on minerals.

① If the collector interacts with the mineral surface by electrostatic force, two conditions must be met: one is that the mineral surface should be positively charged, that is, pH<PZC (or IEP=isoelectric point); another requirement is that most of the reagent itself needs to be dissociated into anions. It can be seen from formula (3-33) that this condition is controlled by controlling $pH > pK_a$, $[A^-] > [HA]$. Therefore, the pH range where the monobasic weak acid flotation reagent can effectively act on minerals by electrostatic force is:

$$pK_a < pH < PZC \tag{3-34}$$

② If the reagent is mainly molecular adsorption on the mineral surface, the pH of the control solution should be: $pH < pK_a, [HA] > [A^-]$.

The K_a values of common anionic flotation reagents are shown in Table 3-5.

Table 3-5 The K_a values of common anionic flotation reagents

Sulfhydryl reagent	K_a	Fatty acid	K_a
Ethyl xanthate①	1.0×10^{-5}	HCOOH	2.1×10^{-5}
Propyl xanthate①	1.0×10^{-5}	CH_3COOH	1.83×10^{-5}
Butyl xanthate①	7.9×10^{-6}	C_2H_5COOH	1.32×10^{-5}
Amyl xanthate①	1.0×10^{-6}	C_3H_7COOH	1.50×10^{-5}
Ethyl dithiophosphate	2.3×10^{-5}	C_4H_9COOH	1.56×10^{-5}
Propyl dithiophosphate①	1.78×10^{-2}	$C_5H_{11}COOH$	1.40×10^{-5}
SN-9	1.6×10^{-7}	$C_6H_{13}COOH$	1.30×10^{-5}
Z-200	3.02×10^{-12}	$C_7H_{15}COOH$	1.41×10^{-5}
n-octyl mercaptan	$10^{-11.8}$	$C_8H_{17}COOH$	1.10×10^{-5}
Hexyl mercaptan	$10^{-6.5}$	$C_{12}H_{23}COOH$	5.10×10^{-6}
Mercaptoacetic acid	1.98×10^{-4}	Oleic acid	1.00×10^{-6}

① measured by M.C.Fuerstenau.

(2) Cationic flotation reagent. Take dodecylamine as an example. There are two forms of dissociation of cations.

Acid dissociation equation:

$$RNH_3^+ \rightleftharpoons RNH_{2(aq)} + H^+ \qquad K_a = \frac{[H^+]\left[RNH_{2(aq)}\right]}{\left[RNH_3^+\right]} = 10^{-10.63} \tag{3-35}$$

Base dissociation equation:

$$RNH_{2(aq)} + H_2O \rightleftharpoons RNH_3^+ + OH^- \qquad K_b = \frac{\left[OH^-\right]\left[RNH_3^+\right]}{\left[RNH_{2(aq)}\right]} = 4.3\times10^{-4} \tag{3-36}$$

The relationship between K_a and K_b is $K_b = K_w / K_a$. The logarithm of equation (3-35) is:

$$pH - pK_a = \lg\frac{\left[RNH_{2(aq)}\right]}{\left[RNH_3^+\right]} \tag{3-37}$$

The flotation significance of the above formula is:

① Electrostatic action: when the mineral surface is negatively charged (pH>PZC) and $pH < pK_a$, $\left[RNH_3^+\right] > \left[RNH_{2(aq)}\right]$, the electrostatic interaction between cationic collector and mineral surface is the most effective, and the floatability of minerals is the largest, such as the flotation of quartz, feldspar and corundum. That is, the effective pH range of electrostatic interaction between cationic collector and mineral is:

$$\mathrm{p}K_{\mathrm{a}} > \mathrm{pH} > \mathrm{PZC} \tag{3-38}$$

② Form molecular complexes. For some minerals, such as zinc oxide ore, the pH value of dodecylamine flotation after vulcanization should reach 10.5-11.5, that is, $\mathrm{pH} > \mathrm{p}K_{\mathrm{a}}$, $[\mathrm{RNH}_{2(aq)}] > [\mathrm{RNH}_3^+]$. At this time, there is another action mechanism, that is, dodecylamine molecule forms a chemical complex $\mathrm{Zn}(\mathrm{RN}_2)_n^{2+}$ with Zn through the arc pair electrons on its N atom.

The K_a values of common cationic collectors are shown in Table 3-6.

Table 3-6 The K_a values of common cationic collectors

Aliphatic amine	K_a	Aliphatic amine	K_a
Nonylamine	1.0×10^{-5}	Pentadecanamine	2.1×10^{-5}
Decamine	1.0×10^{-5}	Cetylamine	1.83×10^{-5}
Undecylamine	7.9×10^{-6}	Octadecylamine	1.32×10^{-5}
Dodecylamine	1.0×10^{-6}	Cetylpyridinium bromide	1.50×10^{-5}
Tridecyl amine	2.3×10^{-5}	*N*-methyldodecylamine	1.56×10^{-5}
Tetradecylamine	1.78×10^{-2}	Dimethyl dodecylamine	1.40×10^{-5}

(3) Amphoteric collector. The dissociation of amphoteric collectors is discussed as follows using amino acids as an example.

Anionic in alkaline solution:

$$\mathrm{RNHCH(CH_3)CH_2COOH + OH^- \rightleftharpoons RNHCH(CH_3)CH_2COO^- + H_2O}$$

Cationic in acidic solution:

$$\mathrm{RNHCH(CH_3)CH_2COOH + H^+ \rightleftharpoons RNH_2^+CH(CH_3)CH_2COOH}$$

The pH in the positive and negative equilibrium state is called the zero electric point pH, which is expressed in $\mathrm{pH_0}$. Experiments show that at $\mathrm{pH_0}$, the behavior of reagent on mineral flotation turns.

① When $\mathrm{pH_0} < \mathrm{PZC}$, the effective condition for electrostatic interaction with minerals is $\mathrm{pH_0} < \mathrm{pH} < \mathrm{PZC}$.

② When $\mathrm{pH_0} > \mathrm{PZC}$, the effective condition is $\mathrm{pH_0} > \mathrm{pH} > \mathrm{PZC}$.

The $\mathrm{pH_0}$ values of some amphoteric collectors are shown in Table 3-7. See Table 7-1 in Chapter 7 for PZC of common minerals.

Table 3-7 The $\mathrm{pH_0}$ values of some amphoteric collectors

Reagent	Zero electric point $\mathrm{pH_0}$
Cetylamino acetic acid $\mathrm{C_{16}H_{33}NHCH_2COOH}$	4.5
N-coconut oil- β- aminobutyric acid $\mathrm{RNHCH(CH_3)CH_2COOH}$	4.1

(continued)

Reagent		Zero electric point pH_0
N-dodecyl- β- aminopropionic acid	$C_{12}H_{25}NHCH_2CH_2COOH$	4.3
N-dodecyl- β- iminodipropionic acid	$C_{12}H_{25}N(CH_2CH_2COOH)_2$	3.7
N-tetradecyl aminoethyl sulfonic acid	$C_{14}H_{29}NHCH_2CH_2SO_3H$	1.0
Stearic acid amino carbonate		6.3-6.6
Stearic aminosulfonic acid		6.3-6.6

3.1.5 The diagram method of solution equilibria of reagent

The previous section discusses the dissociation equilibrium of simple monobasic acids, bases and their salts. For polybasic acids, bases and their salts, such as Na_2S, Na_2CO_3, $H_2C_2O_4$, $CH_3C_6H_4AsO(OH)_2$, the calculation of solution equilibrium is more complex. Therefore, in order to make the calculation results of flotation agent solution balance and its relationship with flotation more simple and clear, various graphic methods are adopted.

This section discusses the diagram of the distribution coefficient (ϕ) of the flotation reagent as a function of pH (called the ϕ-pH diagram method) and the diagram of the variation of the concentration C of each component with the pH value (called the logarithmic diagram of the concentration), namely the lg$C$-pH diagram law.

(1) ϕ-pH diagram. According to Equation (3-22), the side reaction coefficient is defined as

$$a_{\mathrm{A}}=\frac{[\mathrm{A}]'}{[\mathrm{A}]}=1+\beta_1^{\mathrm{H}}\left[\mathrm{H}^+\right]+\beta_2^{\mathrm{H}}\left[\mathrm{H}^+\right]^2+\cdots+\beta_n^{\mathrm{H}}\left[\mathrm{H}^+\right]^n \tag{3-39}$$

The fraction of each component in the solution to the total concentration of A is

$$\frac{[\mathrm{A}]}{[\mathrm{A}]'},\frac{[\mathrm{HA}]}{[\mathrm{A}]'},\frac{[\mathrm{H_2A}]}{[\mathrm{A}]'},\cdots,\frac{[\mathrm{H}_n\mathrm{A}]}{[\mathrm{A}]'}$$

So the distribution coefficient is defined as:

$$\phi_0=\frac{[\mathrm{A}]}{[\mathrm{A}]'}=\frac{1}{a_{\mathrm{A}}}=\frac{1}{1+\beta_1^{\mathrm{H}}\left[\mathrm{H}^+\right]+\beta_2^{\mathrm{H}}\left[\mathrm{H}^+\right]^2+\cdots+\beta_n^{\mathrm{H}}\left[\mathrm{H}^+\right]^n}$$

$$\phi_1=\frac{[\mathrm{HA}]}{[\mathrm{A}]'}=\frac{K_1^{\mathrm{H}}\left[\mathrm{H}^+\right][\mathrm{A}]}{[\mathrm{A}]'}=K_1^{\mathrm{H}}\phi_0\left[\mathrm{H}^+\right]$$

$$\phi_2=\frac{[\mathrm{H_2A}]}{[\mathrm{A}]'}=\frac{\beta_2^{\mathrm{H}}\left[\mathrm{H}^+\right]^2[\mathrm{A}]}{[\mathrm{A}]'}=\beta_2^{\mathrm{H}}\phi_0\left[\mathrm{H}^+\right]^2$$

$$\vdots$$

$$\phi_n = \frac{[H_nA]}{[A]'} = \frac{\beta_n^H[H^+]^n[A]}{[A]'} = \beta_n^H \phi_0 [H^+]^n \tag{3-40}$$

In the formula, ϕ_0, ϕ_1, ϕ_2, ⋯, ϕ_n are the distribution coefficients of each component, indicating the fraction of the concentration of each component in the total concentration. Its value is related to pH. The component distribution diagram of a series of flotation reagents can be calculated from the data of proton constants in Table 3-5 and Table 3-6 (ϕ-PH diagram). The following is an example.

① The ϕ-pH diagram of Na_2S solution. Na_2S is a commonly used vulcanizing reagent for copper oxide, lead, and zinc ore and an inhibitor depressant for sulfide ore. It first undergoes a hydrolysis reaction in solution, and then dissociates:

$$Na_2S + 2H_2O \rightleftharpoons H_2S + 2NaOH \tag{3-41}$$

$$S^{2-} + H^+ \rightleftharpoons HS^- \qquad K_1^H = \frac{[HS^-]}{[S^{2-}][H^+]} = 10^{13.9} \tag{3-42}$$

$$HS^- + H^+ \rightleftharpoons H_2S \qquad K_2^H = \frac{[H_2S]}{[HS^-][H^+]} = 10^{7.02}, \beta_2^H = 10^{20.92} \tag{3-43}$$

$$[S]' = [S^{2-}] + [HS^-] + [H_2S] \tag{3-44}$$

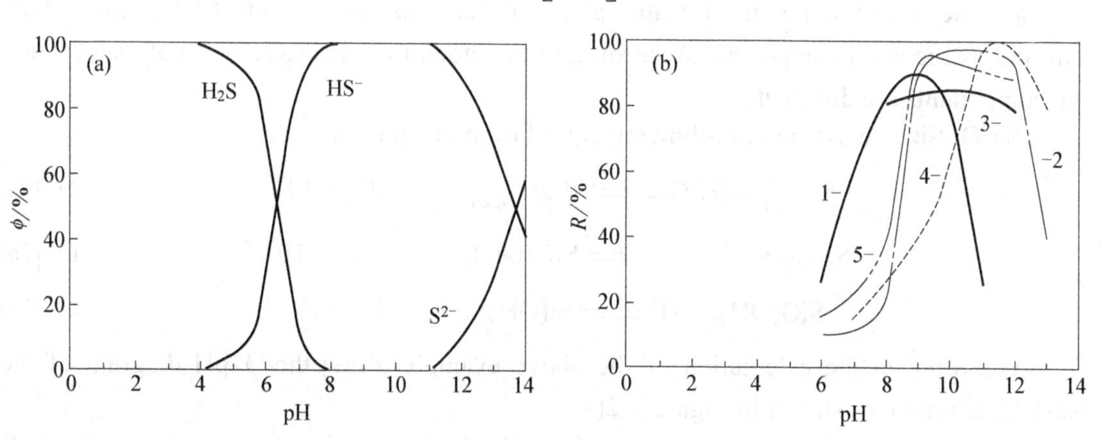

Figure 3-1 Distribution coefficient of S-containing component (a) and the relationship between flotation recovery of copper oxide, lead and zinc and pH in Na_2S solution (b)

1—Malachite Na_2S 50mg/L, BuX(butyl xanthate) 10mg/L; 2—Cerussite Na_2S adjust pH, dodecylamine 4.9mg/L; 3—Cerussite Na_2S 160g/t, AX(amyl xanthate) 80g/t; 4—Hemimorphite Na_2S adjust pH, dodecylamine 2×10^{-5}mol/L; 5—Magnetite after sulfidization dodecylamine 2×10^{-5}mol/L

$$\phi_0 = \frac{[S^{2-}]}{[S]'} = \frac{1}{1 + K_1^H[H^+] + \beta_2^H[H^+]^2} = \frac{1}{1 + 10^{13.9}[H^+] + 10^{20.92}[H^+]^2}$$

$$\phi_1 = \frac{[HS^-]}{[S]'} = K_1^H \phi_0 [H^+] = 10^{13.9} \phi_0 [H^+]$$

$$\phi_2 = \frac{[H_2S]}{[S]'} = \beta_2^H \phi_0 [H^+] = 10^{20.92} \phi_0 [H^+] \tag{3-45}$$

From the equation (3-45), the relationship between the distribution coefficient of each component containing S and pH can be calculated, as shown in Figure 3-1 (a). It can be seen that pH<7.0, H_2S is dominant; pH>7.0, HS^- is the dominant component; pH$>$13.9, S^{2-} is the dominant component. Comparing the flotation results in Figure 3-1(b), it can be found that the optimal pH range of sulfide flotation of oxidized Cu, Pb and Zn ores is in the pH range of HS^- dominant components. Therefore, it is considered that the mechanism of flotation of copper oxide, lead and zinc ores activated by Na_2S is that HS^- reacts with metal atoms on the mineral surface to form sulfide film.

For example, the sulfurization reaction of smithsonite is:

$$ZnCO_{3(surf)} + HS^- \rightleftharpoons ZnS_{surf} + HCO_3^- \tag{3-46a}$$

$$Zn(OH)_{2(surf)} + HS^- \rightleftharpoons ZnS_{surf} + H_2O + OH^- \tag{3-46b}$$

This reaction mechanism has been verified by experiments.

② The ϕ-pH diagram of water glass. Water glass is an industrial product with Na_2SiO_3 as the main component. According to the modulus m ($Na_2O:SiO_2$) of water glass, the components are different.

$Na_2O:SiO_2$ exists in the following equilibrium in solution.

$$SiO_{2(s)} + 2H_2O \rightleftharpoons Si(OH)_{4(aq)} \qquad K_{so} = 10^{-2.7} \tag{3-47a}$$

$$SiO_2(OH)_2^{2-} + H^+ \rightleftharpoons SiO(OH)_3^- \qquad K_1^H = 10^{12.58} \tag{3-47b}$$

$$SiO(OH)_3^- + H^+ \rightleftharpoons Si(OH)_4 \qquad K_2^H = 10^{9.43} \tag{3-47c}$$

According to the calculation of the above example, draw the ϕ-pH diagram of the Na_2SiO_3 solution, as shown in Figure 3-2(a).

It can be seen that when pH<9.4, $Si(OH)_4$ is the dominant component; pH⩾9.4, $SiO(OH)_3^-$ is dominant; pH⩾12.6, $SiO_2(OH)_2^{2-}$ is dominant.

Compared with the results in Figure 3-2(b), it can be seen that the pH value when water glass begins to inhibit fluorite and calcite corresponds to the pH value at which $SiO(OH)_3^-$ begins to form. Accordingly, it should be considered that $SiO(OH)_3^-$ is an effective component that plays an inhibitory role.

③ The ϕ-pH diagram of toluic acid

Toluic acid is a dibasic acid collector and dissociates as follows:

$$CH_3C_6H_4AsO(OH)_2 \rightleftharpoons CH_3C_6H_4AsO(OH)O^- + H^+ \quad K_{a1} = 2.0\times10^{-4} \tag{3-48a}$$

$$CH_3C_6H_4AsO(OH)O^- \rightleftharpoons CH_3C_6H_4AsO_3^{2-} + H^+ \qquad K_{a2} = 2.1\times10^{-8} \qquad (3\text{-}48b)$$

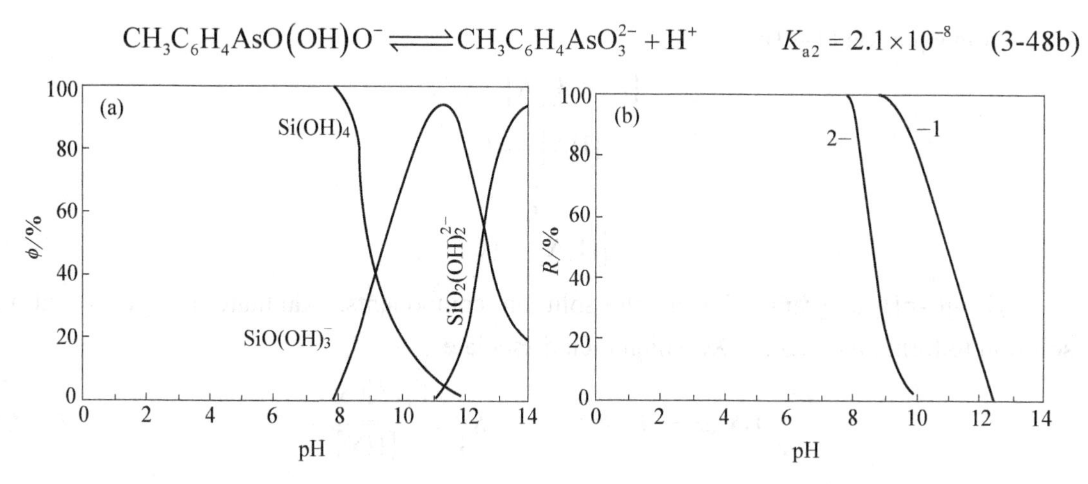

Figure 3-2 The distribution coefficient of water glass components (a), the relationship between recovery rate and pH value (b)

1—Fluorite, Na_2SiO_3, 10^{-4} mol/L; 2—Calcite, Na_2SiO_3, 5×10^{-4} mol/L, [HOl]= 10^{-4} mol/L (HOl=oleic acid)

The composition distribution of toluic acid is drawn as shown in Fig.3-3(a). It can be seen that the optimum pH range of flotation of Pb^{2+} activated quartz and wolframite with toluidine arsonic acid is consistent with the distribution curve of $CH_3C_6H_4AsO(OH)O^-$, and it is inferred that the active component of toluidine arsonic acid is $CH_3C_6H_4AsO(OH)O^-$, which forms a coordination complex with metal ions:

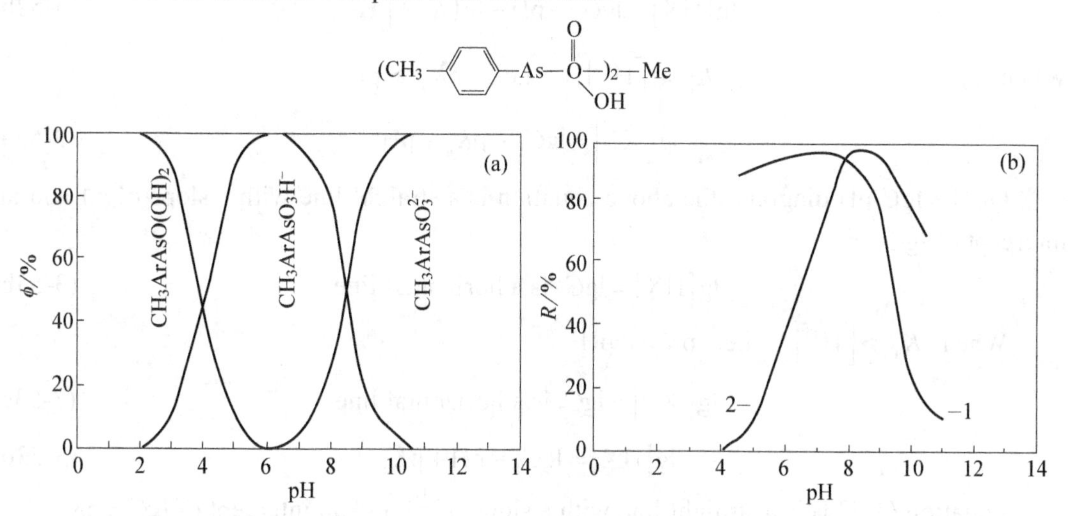

Figure 3-3 The distribution coefficient of toluic acid components (a), the relationship between recovery rate and pH value (b)

Toluidine arsonic acid 50mg/L; $Pb(NO_3)_2$ 10^{-4}mol/L;

1—Wolframite; 2—Quartz

(2) Logarithmic concentration diagram (lg$C$-pH). By introducing the original concentration C_T of the added reagent, the absolute concentration of the components that exist in a certain state of the reagent can be calculated.

From equation (3-40):

$$[A]=\phi_0[A]=\phi_0 C_T$$

$$[HA]=\phi_1 C_T$$

$$\vdots$$

$$[H_nA]=\phi_n C_T \tag{3-49}$$

① $\lg C$-pH diagram of xanthate solution components. Xanthate is hydrolyzed in solution to form xanthogen HX, which then dissociates:

$$HX \rightleftharpoons H^+ + X^- \qquad K_a=\frac{[H^+][X^-]}{[HX]} \tag{3-50}$$

$$\phi_0=\frac{1}{1+K_a^{-1}[H^+]} \qquad [X^-]=\frac{C_T}{1+K_a^{-1}[H^+]} \tag{3-51a}$$

$$\phi_1=K_a^{-1}[H^+]\phi_0 \qquad [HX]=\frac{C_T[H^+]}{K_a+[H^+]} \tag{3-51b}$$

The logarithm of the above two formulas is:

$$\lg[X^-]=\lg C_T-\lg(K_a+[H^+])+\lg K_a \tag{3-52a}$$

$$\lg[HX]=\lg C_T-pH-\lg(K_a+[H^+]) \tag{3-52b}$$

When $K_a \ll [H^+]$, i.e. $pK_a \gg pH$

$$\lg[X^-]=\lg C_T-pK_a+pH \tag{3-53a}$$

On the $\lg C$-pH diagram, the above equation is a straight line with a slope of +1 and an intercept of $\lg C_T - pK_a$.

$$\lg[HX]=\lg C_T \text{ is a horizontal line} \tag{3-53b}$$

When $K_a \gg [H^+]$, i.e. $pK_a \ll pH$

$$\lg[X^-]=\lg C_T \text{ is a horizontal line} \tag{3-53c}$$

$$\lg[HX]=\lg C_T-pH+pK_a \tag{3-53d}$$

Equation (3-53d) is a straight line with a slope of −1 and an intercept of $\lg C_T + pK_a$.

When $K_a=[H^+]$, i.e. $pK_a=pH$

$$\lg[X^-]=\lg[HX]=\lg C_T-\lg 2 \tag{3-54}$$

Therefore, the logarithm diagram of xanthate concentration can be drawn from Equations (3-52) to (3-54), as shown in Figure 3-4.

Significance for flotation: (i) In the study of the flotation mechanism of xanthate, the hypothesis of molecular adsorption and ion adsorption has been proposed. According to the molecular adsorption hypothesis, HX is the effective form of action, then from Figure 3-4, the flotation pH should be at pH<pK_a, such as the flotation of sphalerite (Figure 3-5). However, Figure 3-4 cannot explain the reason why sphalerite does not float after pH<1, although 100% of xanthate is HX.

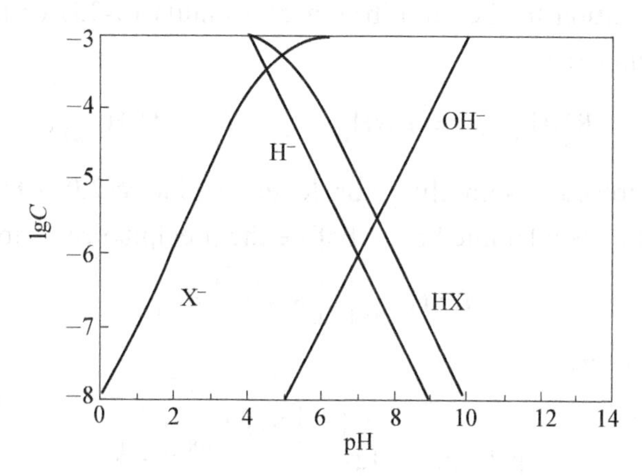

Figure 3-4 The logarithm diagram of xanthate concentration (1.0×10^{-8} mol/L)

According to the ion adsorption hypothesis, the pH of flotation should be pH>pK_a, but it can also be seen from Figure 3-4 that there is competitive adsorption of OH^- at high pH, which may be one of the reasons for the decline of floatability of galena at high pH (see Figure 3-5).

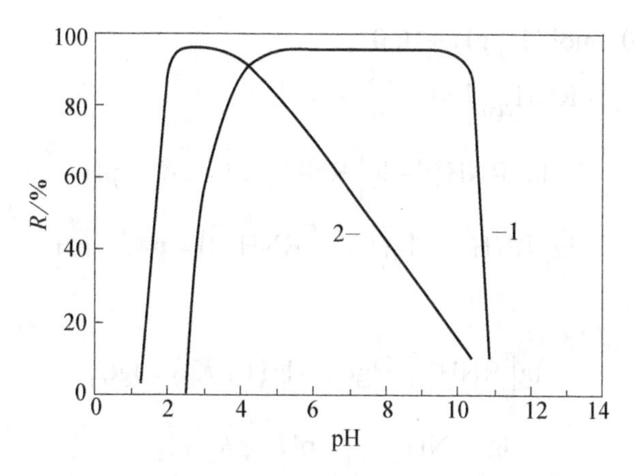

Figure 3-5 The relationship between flotation recovery rate of lead and zinc sulfide ores and pH value

1—galena, BuX 10^{-5} mol/L; 2—sphalerite, AX 2.5×10^{-4} mol/L

(ii) When the original concentration of xanthate is very low, the effect of reagent on flotation is affected by the dissociation of reagent. In addition, it should be noted that the reagent can be significantly effective on flotation when the absolute concentration of reagent is more than 10^{-5} mol/L. If X^- is the effective component, it is required:

$$\lg C_T - \lg\left(K_a + \left[H^+\right]\right) + \lg K_a > 10^{-5}$$

If HX is the effective component, it is required:

$$\lg C_T - pH - \lg\left(K_a + \left[H^+\right]\right) > 10^{-5}$$

② lg*C*-pH diagram of primary amine solution. In the saturated solution of long-chain primary amine, in addition to the equilibrium of formula (3-35) or (3-36), the following dissolution equilibrium exists:

$$RNH_{2(s)} \rightleftharpoons RNH_{2(aq)} \qquad S = \left[RNH_{2(aq)}\right] \tag{3-55}$$

S is called the molecular solubility. For dodecylamine $S=2.0\times10^{-5}$ mol/L, let the initial total concentration of dodecylamine be C_T, before the precipitation is formed:

$$\left[RNH_{2(aq)}\right] + \left[RNH_3^+\right] = C_T \tag{3-56}$$

From equation (3-37):

$$pH - pK_a = \lg\left\{\frac{\left[RNH_{2(aq)}\right]}{C_T - \left[RNH_{2(aq)}\right]}\right\} \tag{3-57}$$

Then the critical pH_s for the formation of precipitation is determined by the following formula:

$$pH_s = pK_a + \lg\left(\frac{S}{C_T - S}\right) \tag{3-58}$$

Set $C_T = 1.0\times10^{-4}\,\text{mol/L}$, $pH_s=10.0$

When $pH < pH_s$, $\left[RNH_{2(s)}\right] = 0$

$$\lg\left[RNH_3^+\right] - \lg\left[RNH_{2(aq)}\right] = pK_a - pH \tag{3-59}$$

$$\lg\left[RNH_3^+\right] - \lg\left(C_T - \left[RNH_3^+\right]\right) = pK_a - pH$$

if $pH \ll pK_a$, then

$$\lg\left[RNH_3^+\right] = \lg C_T - \lg\left(1 + K_a\right) = \lg C_T \tag{3-60a}$$

$$\lg\left[RNH_{2(aq)}\right] = pH - pK_a + \lg C_T \tag{3-60b}$$

When $pH > pH_s$,

$$\left[RNH_{2(aq)}\right] = S = 2.0\times10^{-5} \tag{3-61a}$$

$$\lg\left[RNH_3^+\right] = \lg S - pH + pK_a \tag{3-61b}$$

$$\lg\left[RNH_{2(s)}\right] = \lg\left(C_T - \left[RNH_3^+\right] - S\right) \tag{3-61c}$$

When $pH = pH_s$, from equation (3-59),

$$\lg\left\{\frac{[RNH_3^+]}{[RNH_{2(aq)}]}\right\}=0$$

$$[RNH_3^+]=[RNH_{2(aq)}]=S \tag{3-61d}$$

From formulas (3-60) and (3-61), the concentration of each component of dodecylamine can be obtained as a function of pH, as shown in Figure 3-6.

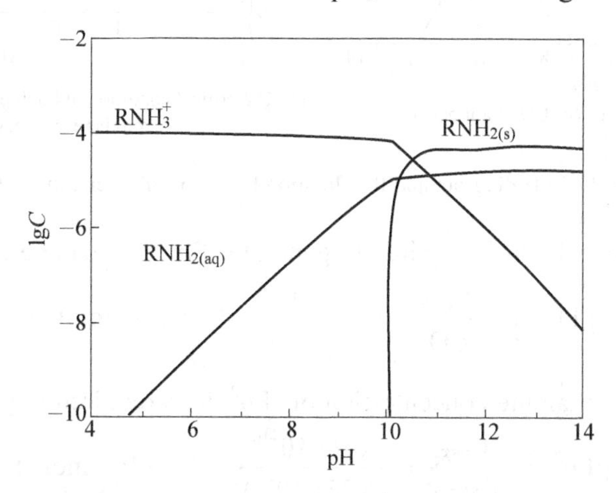

Figure 3-6 Logarithmic concentration diagram for 1×10^{-4} mol/L total dodecylammonium

When the initial concentration of dodecylamine solution is greater than S, dodecylamine precipitation will occur, and the critical precipitation condition pH_s is very important. When $pH>pH_s$, the concentration of RNH_3^+ will decrease sharply, and the concentration of dissolved amine molecules will no longer decrease. Therefore, the upper limit of the effective pH range for the electrostatic effect of dodecylamine on minerals in the previous chapter is $pH<pK_a$. It should be $pH<pH_s$, that is, $pH\leqslant pK_a+\lg S-\lg(C_T-S)$.

③ lg$C$-pH diagram of Na_2CO_3 solution. According to the data of equations (3-2) to (3-5), the lg$C$-pH diagram of Na_2CO_3 is drawn according to the same calculation as above, as shown in Figure 3-7(a). It can be seen that when pH = 6-10, HCO_3^- is the dominant component, and when pH is greater than 10, CO_3^{2-} is the dominant component.

According to Figure 3-7(a), the effective component concentration of Na_2CO_3 on flotation under different pH conditions can be seen. Figure 3-7(b) shows the flotation behavior of galena in Na_2CO_3-EX system (by measuring whether galena is in contact with bubbles). When the EX concentration is 100mg/L (about 6.25×10^{-4} mol/L) and pH>10, cerussite will no longer contact with bubbles. At this time, the concentration of Na_2CO_3 is $C>10^{-4}$ mol/L, and the dominant component is CO_3^{2-}. According to the mechanism model of CO_3^{2-} and X^- competing with Pb^{2+} on the mineral surface:

$$Pb^{2+}+CO_3^{2+}\rightleftharpoons PbCO_3 \qquad L_{PbCO_3}=[Pb^{2+}][CO_3^{2-}]=1.5\times10^{-13} \tag{3-62a}$$

$$Pb^{2+}+2X^-\rightleftharpoons PbX_2 \qquad L_{PbX_2}=[Pb^{2+}][X^-]^2=6.7\times10^{-16} \tag{3-62b}$$

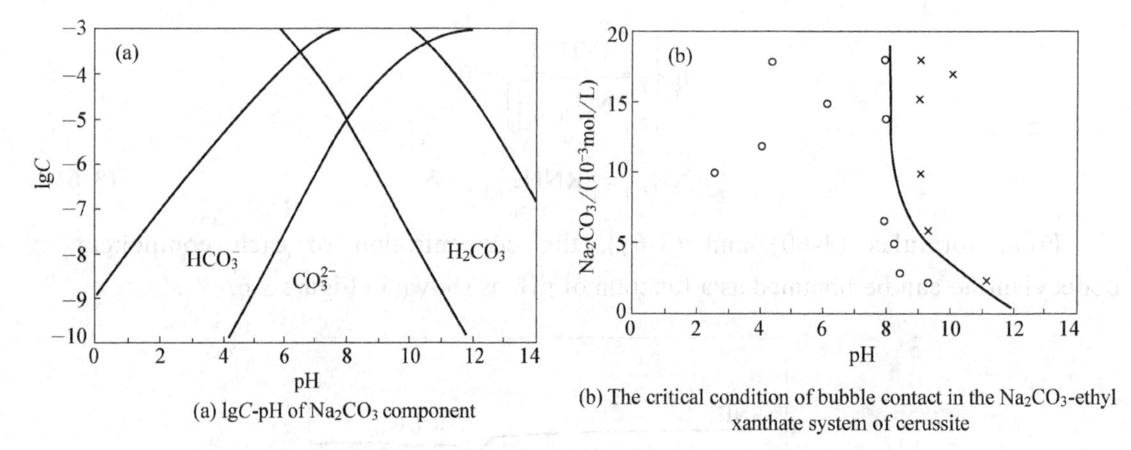

(a) lg$C$-pH of Na_2CO_3 component

(b) The critical condition of bubble contact in the Na_2CO_3-ethyl xanthate system of cerussite

Figure 3-7 The lg$C$-pH diagram and the flotation behavior of galena in Na_2CO_3-EX system

The concentration of Pb^{2+} required to generate $PbCO_{3(s)}$ can be calculated.

$$\left[Pb^{2+}\right]=\frac{L_{PbCO_3}}{\left[CO_3^{2-}\right]}=1.5\times\frac{10^{-13}}{10^{-4}}=1.5\times10^{-9}\,mol/L$$

This value is less than the concentration of Pb^{2+} ions required to generate PbX_2:

$$\left[Pb^{2+}\right]=\frac{L_{PbX_2}}{\left[X^-\right]^2}=\frac{6.7\times10^{-16}}{(6.25\times10^{-4})^2}=1.7\times10^{-9}\,mol/L$$

That is, the formation of $PbCO_3$ is easier than that of PbX_2, indicating that CO_3^{2-} can prevent the reaction of xanthate with minerals.

④ lg$C$-pH diagram of phosphate. Sodium orthophosphate, Na_3PO_4, is hydrolyzed to H_3PO_4 in solution, and then ionized in steps:

$$H_3PO_4 \rightleftharpoons H^+ + H_2PO_4^- \qquad K_{a1}=\frac{1}{K_3^H}=\frac{\left[H^+\right]\left[H_2PO_4^-\right]}{\left[H_3PO_4\right]}$$

$$H_2PO_4^- \rightleftharpoons H^+ + HPO_4^{2-} \qquad K_{a2}=\frac{1}{K_2^H}=\frac{\left[H^+\right]\left[HPO_4^{2-}\right]}{\left[H_2PO_4^-\right]}$$

$$HPO_4^{2-} \rightleftharpoons H^+ + PO_4^{3-} \qquad K_{a3}=\frac{1}{K_1^H}=\frac{\left[H^+\right]\left[PO_4^{3-}\right]}{\left[HPO_4^{2-}\right]} \tag{3-63}$$

The lg$C$-pH diagram can also be drawn, as shown in Figure 3-8(a). At pH<2.2, H_3PO_4 is the dominant component, while the pH ranges of $H_2PO_4^-$, HPO_4^{2-} and PO_4^{3-} dominant components are 2.2<pH<7.2, 7.2<pH<12.4 and pH>12.4, respectively.

Figure 3-8 (b) shows that when sodium oleate (NaOl) is used as a collector, sodium phosphate has no depression effect on scheelite, but depresses fluorite and calcite at pH<9.5 and pH<10.0, respectively. The PZC of the three minerals is about 1.5 for scheelite, 9.5 for fluorite and 10.0 for calcite. In this way, when pH>1.5, the surface of scheelite is negatively charged. It can be seen from Figure 3-8 (a) that after pH>2.2, $H_2PO_4^-$, HPO_4^{2-} and PO_4^{3-} are dominant. So these three negatively charged ions do not adsorb on the surface of negatively charged scheelite and do not depress scheelite. For fluorite, the surface is positively charged at pH<9.5, and the calcite surface is positively charged at pH<10.0. Therefore, the three negatively charged phosphate components are adsorbed on the fluorite surface at pH<9.5, and adsorbed on the calcite surface at pH<10.0, which shows that phosphate inhibits fluorite and calcite at pH<9.5 and pH<10.0 respectively.

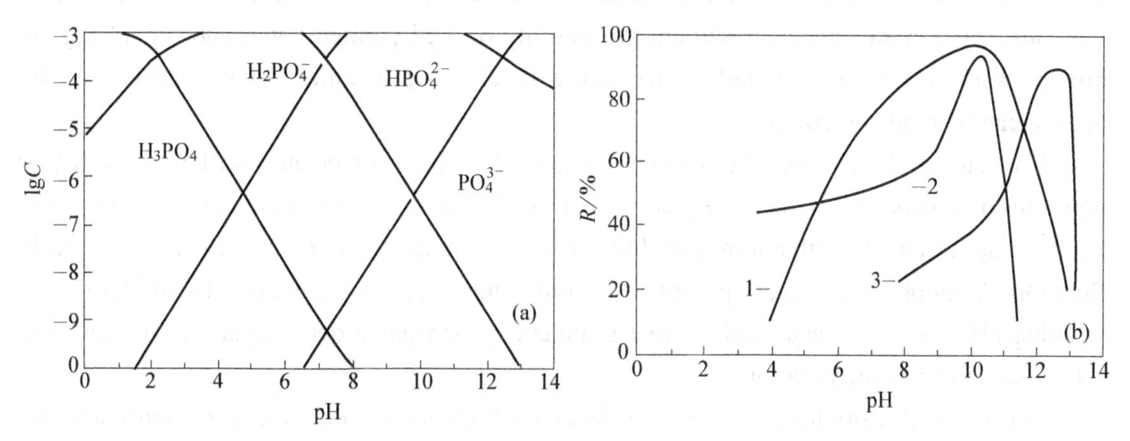

Figure 3-8 Logarithmic distribution diagram of component concentration of phosphate and its adjustment effect on the flotation of scheelite, fluorite and calcite

[NaOl]= 1.31×10^{-3}mol/L; [Na_3PO_4]= 1.58×10^{-3}mol/L; 1—Scheelite; 2—Fluorite; 3—Calcite

Figure 3-9 also shows that there are three turning points in the relationship curve between the adsorption amount of phosphate on goethite and pH, corresponding to pK_{a1}, pK_{a2} and pK_{a3} in Figure 3-8(a). With the increase of the negative charge of the three ions $H_2PO_4^-$, HPO_4^{2-} and PO_4^{3-}, the adsorption capacity decreased rapidly.

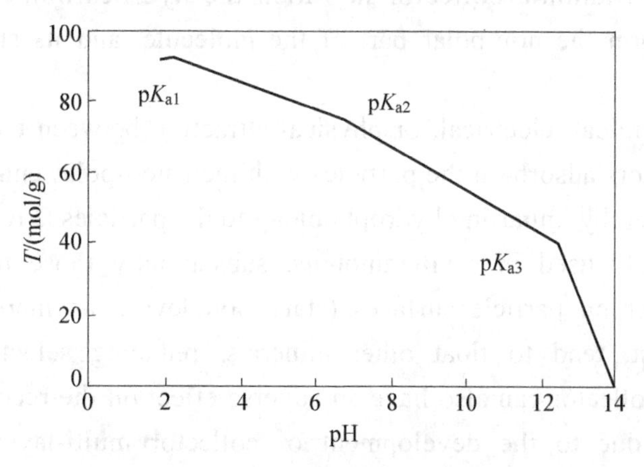

Figure 3-9 Adsorption of phosphate on goethite

3.2 Chemistry of Collector

3.2.1 Collectors and collector salts

Hydrophobicity has to be imparted to most minerals in order to float them. In order to achieve this, surfactants known as collectors are added to the pulp and time is allowed for adsorption agitation in what is known as the conditioning period. Collectors are organic compounds which render selected minerals water-repellent by adsorption of molecules or ions onto the mineral surface, reducing the stability of the hydrated layer, and separating the mineral surface from the air bubble to such a level that attachment of the particle to the bubble can be made on contact.

Collector molecules may be ionising compounds, which dissociate into ions in water, or non-ionising compounds, which are practically insoluble, and render the mineral water-repellent by covering its surface with a thin film. Ionising collectors have found very wide application in flotation. Amphoteric collectors possess a cationic and an anionic function, depending on the working pH, and have been used to treat sedimentary phosphate deposits and to improve the selectivity of cassiterite flotation.

Ionising collectors have complex molecules which are asymmetric in structure and are heteropolar. Collectors are heterogeneous compounds that contain a functional inorganic group coupled with a hydrocarbon chain. In general, the inorganic group is the portion of the collector molecule that adsorbs on the mineral surface while the hydrocarbon chain, being nonionic in nature, provides hydrophobicity to the mineral surface as a result of collector adsorption.

Ionising collectors are classed in accordance with the type of ion, anion or cation that produces the water-repellent effect in water. This classification is given in Figure 3-10. Sodium oleate is an anionic collector in which the hydrocarbon radicals that do not react with water form the non-polar part of the molecule, and its structure is shown in Figure 3-11.

Because of chemical, electrical, or physical attraction between the polar portions and surface sites, collectors adsorb on the particles with their non-polar ends orientated towards the bulk solution, thereby imparting hydrophobicity to the particles (Figure 3-12).

They are usually used in small amounts, substantially those necessary to form a monomolecular layer on particle surfaces (starvation level), as increased concentration, apart from the cost, tend to float other minerals, reducing selectivity. An excessive concentration of a collector can also have an adverse effect on the recovery of the valuable minerals, possibly due to the development of collector multi-layers on the particles, reducing the proportion of hydrocarbon radicals orientated into the bulk solution. The hydrophobicity of the particles is thus reduced, and hence their floatability. The flotation

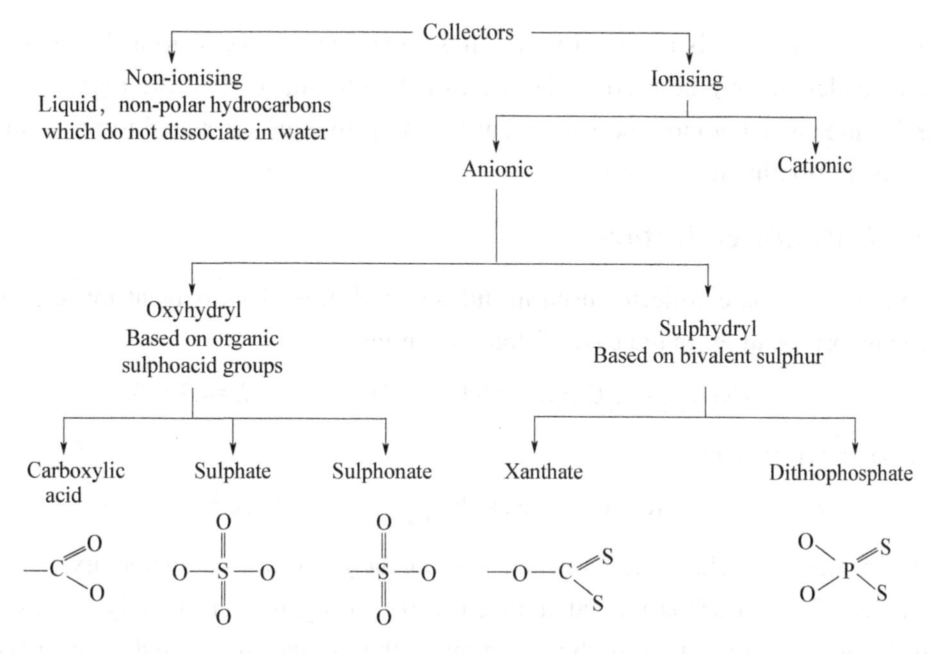

Figure 3-10 Classification of collectors

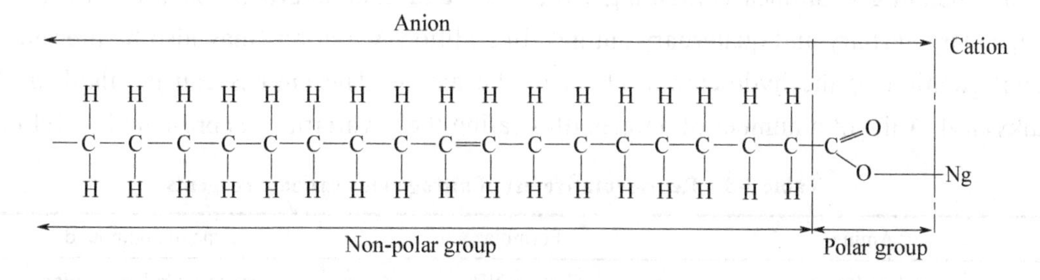

Figure 3-11 Structure of sodium oleate

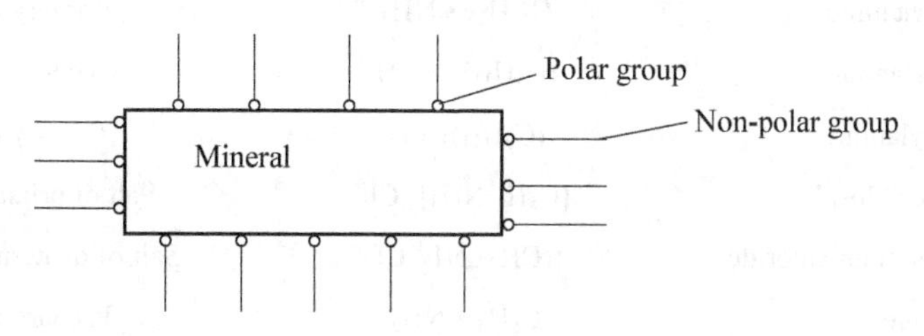

Figure 3-12 Collector adsorption on mineral surface

limit may be extended without loss of selectivity by using a collector with a longer hydrocarbon chain, thus producing greater water repulsion, rather than by increasing the concentration of a shorter chain collector. However, since the solubility of the collector in water rapidly diminishes with increasing chain length and, although there is a corresponding decrease in solubility of the collector products, which therefore adsorb very readily on the mineral surfaces, it is, of course, necessary for the collector to ionize in water for chemisorption to take place on the mineral surfaces. Not only the chain length but also the chain structure, affects solubility and adsorption; branched chains have higher solubility than straight chains. It is common to add more than one collector to a flotation system. A selective collector may be used at the head of the circuit, to float the highly hydrophobic minerals, after which a more powerful, but less selective one, is added to promote recovery of the slower floating minerals.

3.2.2 Cationic collectors

The only cationic collector used in industry is amine. This reagent ionizes in aqueous solution by protonation. In the case of dodecyl-amine:

$$RNH_{2(ad)} + H_2O \rightleftharpoons RNH_3^+ + OH^- \qquad K=4.3\times10^{-4} \tag{3-64}$$

In saturated systems.

$$RNH_{2(s)} \rightleftharpoons RNH_{2(aq)} \qquad K=2\times10^{-5} \tag{3-65}$$

The amines are classified as primary, secondary, tertiary or quaternary depending on the number of hydrocarbon radicals bonded to the nitrogen atom. If only one hydrocarbon group is present with two hydrogen atoms, the amine is termed a primary amine. Correspondingly, amines containing two, three and four hydrocarbon groups are termed secondary, tertiary and quaternary amines. In addition, variations may also be present in the configuration of the hydrocarbon chain of the amine. The amines can be alkyl, aryl and alkylaryl. A list of a number of amines illustrating these variations is presented in Table 3-8.

Table 3-8 Representative list of nitrogenous cationic reagents

Amine	Formula	Type of compound
n-Amylamine	$C_5H_{11} \cdot NH_2$	Primary aliphatic amine
n-Dodecylamine	$C_{12}H_{25} \cdot NH_2$	Primary aliphatic amine
Di-*n*-amylamine	$(C_5H_{11})_2 \cdot NH$	Secondary aliphatic amine
Tri-*n*-amylamine	$(C_5H_{11})_3 \cdot N$	Tertiary aliphatic amine
Amylamine chloride	$[C_5H_{11}NH_3]^+ Cl^-$	Salt of primary aliphatic amine
Tetramethylammonium chloride	$[(CH_3)_4N]^+ Cl^-$	Salt of quaternary ammonium base
Aniline	$C_6H_5 \cdot NH_2$	Primary aromatic amine
p-Toluidine	$CH_3 \cdot C_6H_4 \cdot NH_2$	Primary aromatic amine
Benzylamine	$C_6H_5 \cdot CH_2 \cdot NH_2$	Primary aromatic amine
Diphenylamine	$C_6H_5 \cdot NH \cdot C_6H_5$	Secondary aromatic amine
Alkoxy propylamine acetate	$RO(CH_2)_3NH_2$ ($R=C_8$-C_{10} mixture)	
Technical coco primary amine	RNH_2($R=C_8$-C_{18},55% C_{12})	
Tallow amine acetate	RNH_3Ac	Primary C_{18}

Primary, secondary and tertiary amines are weak bases, whereas quaternary amines are strong bases. Quaternary amines, then, are completely ionized at all values of pH, while ionization of primary, secondary and tertiary amines is pH dependent. Ionization constants of some amines are presented in Table 3-9.

Table 3-9 Ionization constants of various amines in water

Amine	Ionization constant
Methylamine	4.4×10^{-4}
Dimethylamine	5.2×10^{-4}
Trimethylamine	5.5×10^{-4}
Dodecylamine	4.3×10^{-4}
Octadecylamine	4.0×10^{-4}

When 1×10^{-4} mole of tetra methyl ammonium chloride, a quaternary amine, is added to one liter of water, the concentration of $(CH_3)_4N^+$ is 1×10^{-4} mol/L at all values of pH.

On the other hand, when 1×10^{-4} mole dodecylammonium chloride, a primary amine, is added to one liter of water, the concentrations of RNH_3^+, $RNH_{2(aq)}$, $RNH_{2(s)}$ vary with pH. For example, the concentrations of these three species for this addition of amine at pH 8 and pH 12 are presented in Table 3-10.

Table 3-10 The pH$_0$ values of some amphoteric collectors

pH	Species / (mol/L)		
	RNH_3^+	$RNH_{2(aq)}$	$RNH_{2(s)}$
8	9.89×10^{-5}	2.32×10^{-7}	0
12	8.60×10^{-7}	2.00×10^{-5}	7.91×10^{-5}

Concentrations at additional pH values are presented in a logarithmic concentration diagram in Figure 3-6.

(1) Solubility. As the length of the hydrocarbon chain is increased, the solubility of the amine is reduced. The solubilities of the undissociated molecules of decyl, dodecyl and tetradecylamine are listed in Table 3-11. These values represent the concentrations of $RNH_{2(aq)}$ that are present at high values of pH where the concentration of RNH_3^+ is small and where $RNH_{2(s)}$ is present. In the case of dodecylamine, the solubility of $RNH_{2(aq)}$ and RNH_3^+ at pH 13.0 is 2×10^{-5} mol/L, since RNH_3^+ is 8.6×10^{-8} mol/L and $RNH_{2(aq)}$ is 2×10^{-5} mol/L. At pH 8.0, however, the solubility is 8.62×10^{-3} mol/L, i.e. RNH_3^+ is 8.6×10^{-3} mol/L and $RNH_{2(aq)}$ is 2×10^{-3} mol/L.

Table 3-11 Solubilities or undissociated molecules of various amines

Amines	Solubilities / (mol/L)
Decylamine	5×10^{-4}
Dodecylamine	2×10^{-5}
Tetradecylamine	1×10^{-6}

(2) Micellization. Micellization, a characteristic of hydrocarbon chain surfactants, assumes an important role in flotation systems. Micelles are aggregates of collector ions of colloidal size that form by van der Waal's bonding between hydrocarbon chains of the collectors.

Micelles form because the hydrocarbon chain is nonionic in nature, and a mutual incompatibility between polar water molecules and non polar hydrocarbon chains exists. When a certain concentration of surfactant ions is reached in solution, termed the critical micelle concentration, CMC, the hydrocarbon chains associate into aggregates or micelles and come out of solution. These aggregates cannot be seen by eye, but their presence can be noted by their ability to scatter light when a beam is passed through the solution. The free energy decrease that the system experiences when this phenomenon occurs is significant, namely 0.62 kcal/mol per CH_2 unit. For dodecylamine, then, the free energy decrease is about 7.5 kcal/mol when micelles are formed. The critical micelle concentrations of various amines are given in Table 3-12.

Table 3-12 Critical micelle concentrations of various amines

Amines	CMC/(mol/L)
Decylamine	3.2×10^{-2}
Dodecylamine	1.3×10^{-2}
Tetradecylamine	4.1×10^{-3}
Hexadecylamine	8.3×10^{-4}
Octadecylamine	4.0×10^{-4}

As the hydrocarbon chain length is increased, the concentration at which micellization occurs is reduced in accordance with the free energy considerations mentioned previously.

There is a limit to the number of collector ions that can be contained within a micelle due to the electrostatic repulsion between charged heads. The presence of inorganic salts whose ions are counter ions for the collector ions, or neutral organic molecules, such as long-chain alcohols, reduces the repulsive force between the charged heads and, hence, lowers the CMC. For example, with an addition of 5×10^{-2} molar sodium chloride, the CMC of dodecylammonium chloride is reduced from 1.3×10^{-2} to 6.8×10^{-3} molar.

3.2.3 Oxhydryl anionic collectors

The collectors that fall into this category, are the carboxylates (fatty acids), sulfonates, alkyl sulfonates, and certain chelating reagents. The structural formulas of these reagents are presented in Table 3-13. Fatty acids dissociate into negatively-charged carboxylate.

$$RCOOH(aq) \rightleftharpoons H^+ + RCOO^- \tag{3-66}$$

Table 3-13 Structural formulas of sodium salts of various anionic collectors

Collector	Structural formula
Carboxylate	$R-C(=O)-O^- \cdots Na^+$
Sulfonate	$R-S(=O)_2-O^- \cdots Na^+$
Alkyl sulfate	$R-O-S(=O)_2-O^- \cdots Na^+$
Hydroxamate	$R-C(=O)-S(=H)(-O^-) \cdots Na^+$

All of the fatty acids are weak acids whose average $pK_a=4.7\pm0.5$. They are available in a variety of molecular weights and configurations. The hydrocarbon chain may be straight and saturated or straight with one, two or three double bonds. In other cases the fatty acid may be comprised of hexagonal ring structures on which other alkyl chains are attached. Hydrocarbon chain configurations and solubilities of the undissociated molecule, $RCOOH_{(aq)}$, are presented in Table 3-14.

Table 3-14 Structure and solubility of selected fatty acides

Fatty acid	Formula	Solubility of undissociated molecule, 20℃/(mol/L)
Capric	$CH_3(NH_2)_8COOH$	3.0×10^{-4}
Lauric	$CH_3(NH_2)_{10}COOH$	1.2×10^{-5}
Myristic	$CH_3(NH_2)_{12}COOH$	1.0×10^{-6}
Palmitic	$CH_3(NH_2)_{14}COOH$	6.0×10^{-7}
Stearic	$CH_3(NH_2)_{16}COOH$	3.0×10^{-7}
Oleic	$CH_3(CH_2)_7CH=CH(CH_2)_7COOH$	5.2×10^{-20}

In commercial practice the two fatty acids that are used most frequently are impure oleic acid and Talloel which consists largely of abietic acid. Since fatty acids are weak acids, the concentrations of carboxylate ion and the molecular species, $RCOOH_{(aq)}$, are dependent on solution pH. For example, when 1×10^{-4} mole of stearic acid is added to one liter of water, the concentrations of $RCOO^-$ and $RCOOH_{(aq)}$ at pH 4 and pH 8 are shown in Table 3-15.

Table 3-15 The pH_0 values of some amphoteric collectors

pH	Species /(mol/L)		
	$RCOO^-$	$RCOOH_{(aq)}$	$RCOOH_{(s)}$
4	6.00×10^{-8}	3.00×10^{-7}	9.96×10^{-5}
8	9.99×10^{-5}	5.00×10^{-8}	0

Hydroxamic acids are weak acids, the pK_a being around 9. On the other hand, sulfonic acid is a strong acid, the pK_a being about 1.5.

(1) Solubility. The solubilities of the undissociated molecules of various fatty acids presented in Table 3-14 are the concentrations of the molecular species at low values of pH where the concentrations of carboxylate ion are small. At pH 2.0, for instance, the solubility of stearic acid is 3×10^{-7} mol/L. That is, the concentration of $RCOO^-$ is 6×10^{-10} mol/L, while $RCOOH_{(aq)}$ is 3×10^{-7} mol/L. At pH 8.0, however, the solubility is 6×10^{-4} mol/L, i.e. $RCOO^-$ is 6×10^{-4} mol/L and $RCOOH_{(aq)}$ is 3×10^{-7}mol/L. The solubility of those species containing double bonds in their hydrocarbon chains, such as oleic acid, is greater than the saturated varieties, such as stearic acid. The double bonds have a polar character, and polar water molecules are attracted to them. The study of oleate solutions has provided evidence which indicates the existence of the oleate dimer $[(RCOO)_2^{2-}]$ and an iono-molecular species ($RCOOH\cdot RCOO^-$) know as an acid soap in addition to oleate ion ($RCOO^-$) and oleic acid (RCOOH). A distribution diagram for an oleate concentration of 3×10^{-5} mol/L (Figure 3-13) has been presented. It is seen in this figure that:

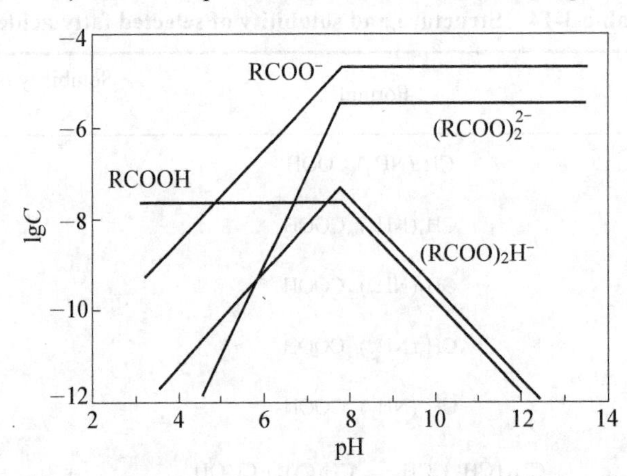

Figure 3-13 Species distribution diagram for an aqueous oleate solution

① The maximum activity of the iono-molecular species ($RCOOH\cdot RCOO^-$) occurs at pH 7.8 for a total oleate concentration of 3×10^{-5} mol/L.

② The activities of the ionic oleate ($RCOO^-$) and the dimer $(RCOO)_2^{2-}$ increase as the pH is increased up to pH 7.8 and then remain constant at higher pH values, and

③ Neutral oleic acid (RCOOH) starts to precipitate from this aqueous solution at pH values less than 7.8.

Zimmels and Lin have studied the stepwise association of soaps in aqueous solutions and have defined the critical micelle concentration as a concentration above which additional surfactant molecules are transformed mainly into associated micelle forms. Conventionally there is only one critical micelle concentration. As can be seen in Figure 3-14, the curve has three sharp changes in slope at 1.7×10^{-4} mol/L, 4.6×10^{-4} mol/L and 2.1×10^{-3} mol/L which have been attributed to three associative forms, the last one being the conventional CMC. Cook as well as Stains and Alexander has quantitatively established the role of premicellar associations and explained the hydrolysis behavior.

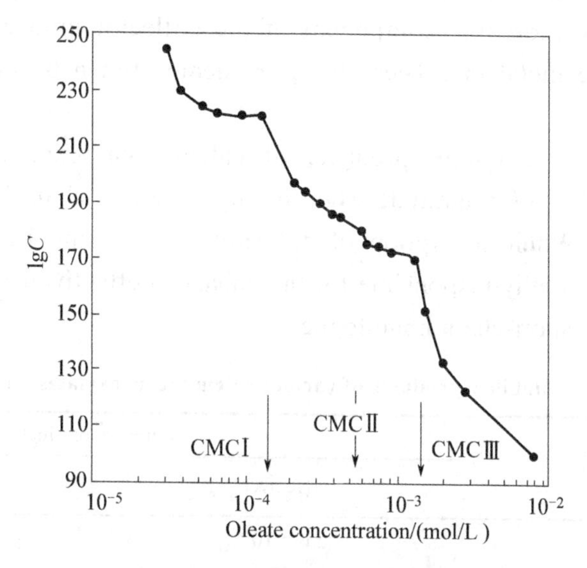

Figure 3-14 Equivalent conductance of sodium oleate solutions as a function of oleate concentration

Sulfonates and alkyl sulfates possess high solubilities in water as shown in Table 3-16. Since alkyl sulfates are salts of strong acids, the dissolved species will be essentially all RSO_4^-.

Table 3-16 Solubilities of various alkyl sulfates in water, 30℃

Alkyl sulfates	Solubilities/(mol/L)
Dodecyl sulfate	5
Tetradecyl sulfate	1×10^{-1}
Hexadecyl sulfate	9×10^{-4}

(2) Micellization. Critical micelle concentrations of various carboxylates, sulfonates and alkyl sulfates are listed in Table 3-17.

Table 3-17 Critical micelle concentrations of various carboxylates, sulfonates and alkyl sulfates

Chain length	CMC/(mol/L)		
	Carboxylate	Sulfonate	Alkyl sulfate
C_{12}	2.6×10^{-2}	9.8×10^{-3}	8.2×10^{-3}
C_{14}	6.9×10^{-3}	2.5×10^{-3}	2.0×10^{-3}
C_{16}	2.1×10^{-3}	7.0×10^{-4}	2.1×10^{-4}
C_{18} ①	1.8×10^{-3}	7.5×10^{-4}	3.0×10^{-4}

① Temperature, 50℃; other determinations at room temperature.

(3) Solubility of collector salts. The solubilities of polyvalent metal-collector salts are a function of the inorganic component of the collector molecule, the hydrocarbon chain length and the metal ion. These facts are demonstrated by the solubility products listed in Table 3-18.

As indicated, the solubility products of calcium carboxylates are decreased by approximately 11 orders of magnitude when the hydrocarbon chain length is increased by eight carbon atoms. When adsorption of collector is occurring by chemisorptions, this phenomenon is substantially responsible for the enhanced effectiveness of the long-chained collectors over that of short-chain homologues.

Table 3-18 Solubility products of various calcium carboxylates and sulfonates

Carbon atoms in molecule	Solubility product	
	$Ca(RCOO)_2$	$Ca(RSO_3)_2$
8	2.69×10^{-7}	—
10	3.80×10^{-10}	8.51×10^{-9}
12	7.94×10^{-13}	4.68×10^{-11}
14	1.00×10^{-15}	2.88×10^{-14}
16	3.80×10^{-18}	1.58×10^{-16}
18(oleate)	1.80×10^{-16}	—

3.2.4 Sulphydryl anionic collectors

The most commonly used sulfhydryl anionic collector is xanthate, but a number of other sulfur-bearing surfactants have also found application. The sodium salts of a number of these collectors are presented in Table 3-19.

Table 3-19 Various anionic sulfhydryl collectors

Collectors	Molecular formula
Xanthate	$R{-}O{-}\underset{\|\|}{\overset{}{C}}{-}S^- \cdots Na^+$ (C=S)

(continued)

Collectors	Molecular formula
Thiophosphate	$(RO)_2P(=S)S^- \cdots Na^+$
Thiocarbamate	$R_2N{-}C{-}S^- \cdots Na^+$
Mercaptan	$R{-}S^- \cdots Na^+$
Thiourea	$R_2N{-}C(=NH){-}S^- \cdots Na^+$
Mercaptobenzothiazole	$C_6H_4(N{=}C(S^-){-}S) \cdots Na^+$

Note: R represents the hydrocarbon chain.

(1) Dissociation constants of xanthic acids. The dissociation constants of the short-chain xanthic acids are presented in Table 3-20.

Table 3-20 Dissociation constants of various xanthic acids

Xanthic acids	Dissociation constant
C_2	0.029
C_3	0.025
C_4	0.023
C_5	0.019

As shown, the xanthic acids are relatively strong acids. In neutral and basic media, then, xanthate will be present as xanthate ion.

(2) Solubility. Alkali metal and alkaline earth xanthates are soluble in water. For example, sodium ethyl xanthate has a solubility of 8 mol/L. On the other hand, heavy-metal xanthate salts possess only limited solubility in aqueous solution. This fact may be seen from the solubility products of several metal xanthates which are listed in Table 3-21.

Table 3-21 Solubility products of various metal xanthates

Metal xanthate	Solubility product
Zinc ethyl xanthate	4.9×10^{-9}
Zinc amyl xanthate	1.6×10^{-12}
Zinc octyl xanthate	1.5×10^{-16}
Lead ethyl xanthate	2.1×10^{-17}
Lead amyl xanthate	1.0×10^{-24}
Cuprous ethyl xanthate	5.2×10^{-20}

Solubility products of the metal xanthates can be noted to decrease with increased chain length. Further, the fact that the solubility product of zinc ethyl xanthate is considerably greater than either lead ethyl xanthate or cuprous ethyl xanthate can also be noted. This phenomenon has important ramifications in the selective flotation separations of sphalerite, chalcopyrite and galena.

Solubility products of metal dithiophosphate salts are greater than those of the corresponding metal xanthates. See Table 3-22.

Table 3-22 Solubility products of various metal dithiophosphates

Metal dithiophosphate	Solubility product
Zinc diethyl dithiophosphate	1.5×10^{-2}
Zinc diamyl dithiophosphate	1.0×10^{-8}
Lead diethyl dithiophosphate	7.5×10^{-12}
Lead diamyl dithiophosphate	4.2×10^{-18}
Cupric diethyl dithiophosphate	1.4×10^{-16}

In the precipitation of metal xanthate salts in solution, the stoichiometry of the metal salt is generally two to one for divalent metal ions. For example, lead ethyl xanthate is $Ph(EX)_{2(S)}$ where EX^- represents ethyl xanthate ion. In the case of cupric ion, however, oxidation of xanthate to dixanthogen by Cu^{2+} occurs during which Cu^{2+} reduction to Cu^+ takes place. Cuprous xanthate, being very insoluble, forms by precipitation. The canary yellow colored precipitate in a xanthate-copper system is cuprous xanthate formed by the following reaction:

$$Cu^{2+} + 2EX^- \longrightarrow CuEX_{(S)} + \frac{1}{2}(EX)_2 \tag{3-67}$$

where EX^- and $(EX)_2$ represent ethyl xanthate and diethyl dixanthogen, respectively. Although cupric xanthate forms initially in this system, it is not a stable species, and its existence is only temporary. By the same token, ferric xanthate is also an unstable species. It is not found as a separate solid phase in an iron-xanthate system, Fe^{3+} oxidizes xanthate to dixanthogen in acid medium. Ferric hydroxide is the stable iron species in basic medium, and xanthate oxidation does not occur in the presence of added ferric salts under these conditions.

(3) Oxidation of sulfhydryl collectors . In addition to oxidation by cupric ion and ferric ion, xanthate can also be oxidized to the dimer, dixanthogen, by dissolved oxygen. The oxidation reactions are:

$$2X^- + \frac{1}{2}(O_2) + H_2O \longrightarrow X_2 + 2OH^- \tag{3-68}$$

$$2X^- + Cu^{2+} \longrightarrow CuX_{(S)} + \frac{1}{2}(X_2) \tag{3-69}$$

$$2X^- + 2Fe^{3+} \longrightarrow 2Fe^{2+} + X_2 \tag{3-70}$$

Although xanthate oxidation by dissolved oxygen is thermodynamically favorable, it is slow kinetically. As a result dixanthogen formation in this manner can be assumed not to occur to any appreciable extent in flotation systems. It should be mentioned however, that electrochemical oxidation of xanthate by oxygen on an appropriate solid surface, such as pyrite, does occur.

With Cu^{2+} and Fe^{3+}, however, reaction is relatively fast kinetically. As shown in Figure 3-15, xanthate is oxidized completely to dixanthogen at pH 2 with Fe^{3+} but is not oxidized at pH 6 and above with a 10-minute reaction period. In the case of Cu^{2+}, essentially complete oxidation of xanthate is effected up to pH 10, while no oxidation is possible at about pH 11 and above with this same reaction time.

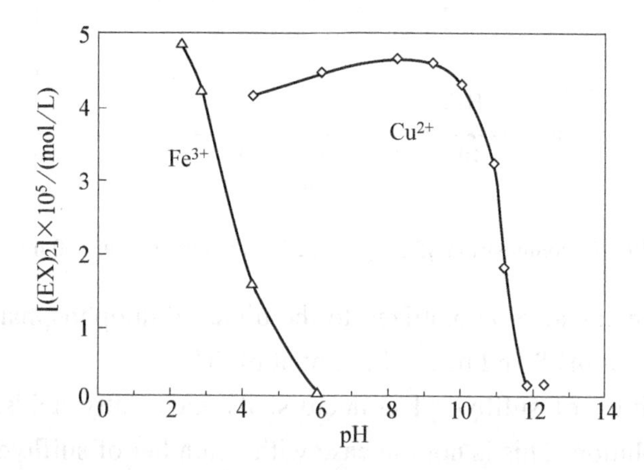

Figure 3-15 Effect of pH on the oxidation of 2×10^{-4} mol/L ethyl xanthate by 1×10^{-4} mol/L cupric or ferric species (10-minute reaction time)

The rate at which dixanthogen forms under these conditions can be seen from the data given in Figure 3-16. Eighty percent of the xanthate is oxidized after two minutes of reaction time when Cu^{2+} is present.

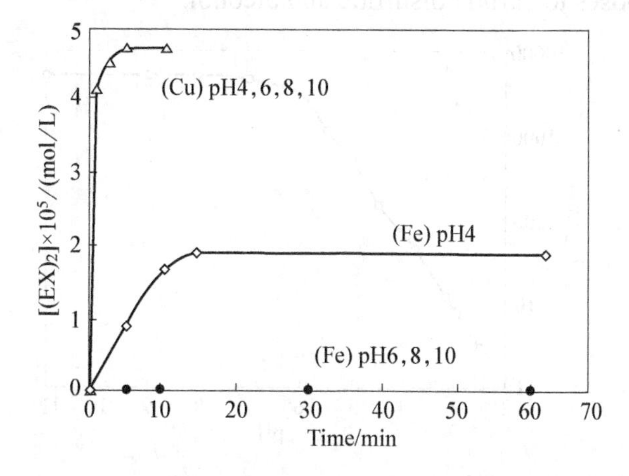

Figure 3-16 Rate of formation of diethyl dixanthogen from the oxidation of 2×10^{-4} mol/L ethyl xanthate by 1×10^{-4} mol/L cupric or ferric species

With regard to dixanthogen stability, importantly, dixanthogen is not stable above about pH 10.5. Xanthate ion is the stable species under these conditions. The rate at which dixanthogen reduces to xanthate ion is presented as a function of pH in Figure 3-17. Decomposition of alkyl dixanthogens has also been studied by Jones et al.

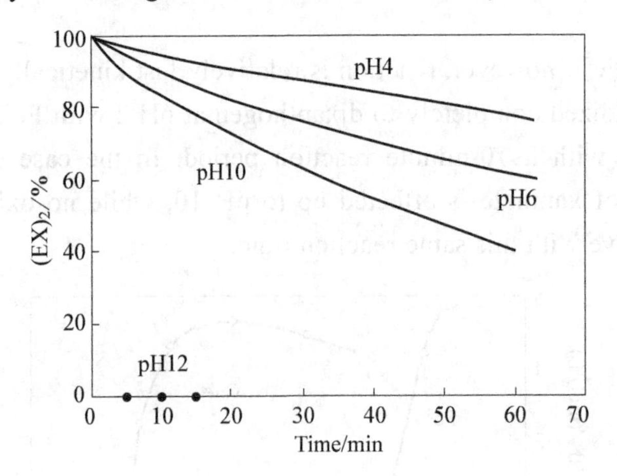

Figure 3-17 Decomposition of 2 mg of diethyl dixanthogen in 100 mL of solution

Dithiophosphate can also be oxidized to the dimer, dithiophosphatogen, by cupric ion. Oxidation is possible at pH 8 and below but not at pH 10.

(4) Decomposition of sulfhydryl collectors. Amines, fatty acids, and sulfonates are stable in aqueous solution. This is not the case with a number of sulfhydryl collectors whose stability is dependent on solution pH (Table 3-23). In the case of ethyl xanthate, decomposition as a function of time and pH is given in Figure 3-18. The half-life of ethyl xanthate is reduced dramatically as the pH is reduced. At pH 2.5, for example, the half-life of ethyl xanthate is 2 minutes. In alkaline media, xanthate ion is not decomposed quickly and can be considered stable toward decomposition in flotation operations. In acid medium xanthic acid decomposes to carbon disulfide and alcohol.

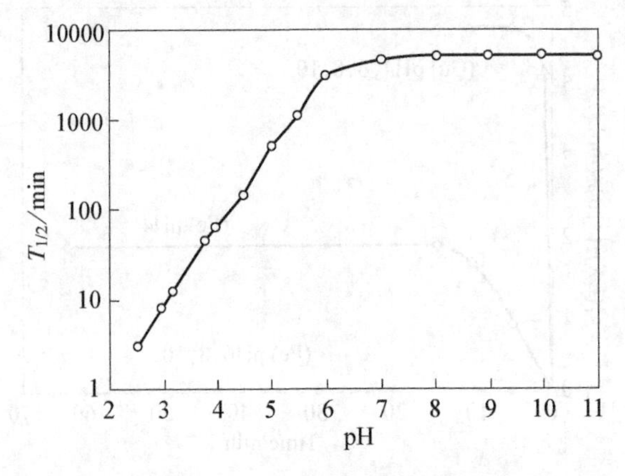

Figure 3-18 Half-life of ethyl xanthate as a function of pH

$$HEX_{(aq)} \longrightarrow CS_2 + ROH \tag{3-71}$$

where $HEX_{(aq)}$ and ROH represent ethyl xanthic acid and ethyl alcohol, respectively.

Xanthates are decomposed slowly in alkaline medium, but the mechanism is different from that in acid medium. Reaction products are carbon disulfide and mono-thiocarbonate.

(5) Hydrolysis of collectors. When salts of weak acids or weak bases are added as collectors, the collector ion may hydrolyze to the weak acid or weak base depending on solution pH.

Table 3-23 Stability ranges of various sulfhydryl collectors

Collector	pH range
Xanthate	8-13
Dixanthogen	1-11
Dithiophosphate	4-12
Dithiocarbamate	5-12
Thionocarbamate	4-9
Mercaptobenzothiazole	4-9

This is the case for most flotation collectors. Salts of quaternary amine, sulfonate and alkyl sulfate do not hydrolyze, however, since their respective bases of acids are strong. The hydrolysis reactions for dodecylammonium ion, oleate ion and ethyl xanthate ion are as follows for additions of dodecylammonium chloride, sodium oleate and sodium ethyl xanthate:

$$RNH_3^+ + H_2O \rightleftharpoons RNH_{2(aq)} + (H_3O)^+ \qquad pK=10.6 \tag{3-72}$$

$$Ol^- + H_2O \rightleftharpoons HOl_{(aq)} + OH^- \qquad pK=9.0 \tag{3-73}$$

$$EX^- + H_2O \rightleftharpoons HEX_{(aq)} + OH^- \qquad pK=12.5 \tag{3-74}$$

The predominant dissolved species of these collectors at various valued of pH are as shown in Table 3-24.

Table 3-24 The predominant dissolved species of these collectors

Collector	Predominant species	
Dodecylamine	Below pH 10.6 RNH_3^+	Above pH 10.6 $RNH_{2(aq)}$
Oleate	Below pH 10.6 $HOl_{(aq)}$	Above pH 5.0 Ol^-
Ethyl xanthate	Below pH 10.6 $HEX_{(aq)}$	Above pH 1.5 EX^-

Chapter 4
Chemistry of Mineral Solution

Minerals have a certain solubility, especially salt minerals, which are more soluble. In their saturated aqueous solution, more mineral lattice ions are dissolved, which will have a greater impact on the flotation process. There are some data on the solubility of minerals, but they are not complete, and there may be differences in the solubility of the same minerals from different origins. According to the thermodynamic data, the theoretical value of the solubility of minerals under certain conditions can be obtained from the stoichiometric formula of minerals and various equilibrium relations, so as to discuss their influence on flotation.

4.1 Solubility of Minerals in Pure Water

Assuming that after the minerals are in equilibrium in water, the dissolved components all exist in the form of ions M^{n+} and A^{m-}, where M^{n+} is the mineral cation and A^{m-} is the mineral anion, the following equilibrium exists in a saturated aqueous solution of minerals:

$$M_mA_n \rightleftharpoons mM^{n+} + nA^{m-} \tag{4-1}$$

The corresponding equilibrium constant is:

$$K_{sp} = [M^{n+}]^m[A^{m-}]^n \tag{4-2}$$

K_{sp} is the solubility product of the mineral.

Since there are often other ligands that can complex with M in the solution, this complexation reaction, the hydrolysis reaction of M^{n+} and the proton addition reaction of A^{m-} will all affect the solubility of minerals. In this case, the concept of the conditional solubility product should be used. The conditional solubility product of a mineral is defined as:

$$K'_{sp} = [M]'^{m}[A]'^{n}$$

Where [M]′ and [A]′ are the total concentrations of M and A in the mineral saturated aqueous solution, respectively.

[M]′ is related to the free metal ion concentration $[M^{n+}]$ by the side reaction coefficient a_M. $[A]'$ is connected with $\left[A^{m-}\right]$ by the side reaction coefficient a_A of the proton addition reaction.

Therefore:

$$a_M = [M]'/[M^{n+}], a_A = [A]'/[A^{m-}] \tag{4-3}$$

$$K'_{sp} = [M^{n+}]^m [A^{m-}]^n a_M^m a_A^n = K_{sp} a_M^m a_A^n \tag{4-4}$$

For MA-type minerals, there is:

$$S_m = k_{sp} a_M a_A \tag{4-5}$$

Let S_m be the solubility of Y mineral, represented by mol/L, $[M]' = mS_m$, $[A]' = nS_m$, then:

$$K'_{sp} = (mS_m)^m (nS_m)^n = K_{sp} a_M^m a_A^n \tag{4-6}$$

$$S_m = \left(\frac{K_{sp} a_M^m a_A^n}{m^m n^n}\right)^{\frac{1}{n+m}} \tag{4-7}$$

For MA-type minerals, there is:

$$S_m = (K_{sp} a_M a_A)^{1/2} \tag{4-8}$$

Due to the differences in the hydrolysis and complexation reactions of cations and the proton addition reaction of anions in different minerals, the calculation of the solubility of sulfide minerals, oxide minerals and salt minerals is discussed below.

4.1.1 Solubility of sulfide minerals

When calculating the solubility of sulfide ores, the hydrolysis reaction of mineral cations and the proton addition reaction of anion S^{2-} should be considered:

In a saturated aqueous solution of sulfide ore, the total concentration of S is:

$$[S'] = [S^{2+}] + [HS^-] + [H_2S] = [S^{2+}]\left(1 + K_1^H[H^+] + \beta_2^H[H^+]^2\right)$$

$$a_s = 1 + K_1^H[H^+] + \beta_2^H[H^+]^2 \tag{4-9}$$

The total concentration of metal ions is:

$$[M]' = [M^{n+}] + \left[M(OH)^{n-1}\right] + \cdots + \left[M(OH)_k^{n-k}\right]$$

$$a_M = 1 + K_1[OH^-] + \beta_2[OH^-]^2 + \cdots + \beta_K[OH^-]^K \tag{4-10}$$

To M_mS_n,

$$K'_{sp} = K_{sp} a_M^m a_S^n \tag{4-11}$$

For the example of galena:

$$PbS \rightleftharpoons Pb^{2+} + S^{2-} \qquad K_{sp} = 10^{-27.5} \tag{4-12}$$

Since the solubility product of PbS is very small, the pH change of the solution caused by the addition of protons to S^{2-} can be ignored, and the pH of the solution is regarded as 7, then:

$$a_S = 1 + 10^{13.9}[H^+] + 10^{20.92}[H^+]^2 = 1.63 \times 10^7$$

$$a_{Pb} = 1 + 10^{6.3}[OH^-] + 10^{10.9}[OH^-]^2 + 10^{13.9}[OH^-]^3 = 1.2$$

$$K'_{sp} = 10^{-27.5} \times 1.63 \times 10^7 \times 1.2 = 6.19 \times 10^{-21}$$

$$S_m = (K'_{sp})^{1/2} = 7.87 \times 10^{-11}$$

Similarly, the solubility of other sulfide ores in pure water can be calculated from the data in Tables 4-1 and 4-2. The calculation results are shown in Table 4-3. It can be seen that the solubility of sulfide ores in pure water is generally very small. The data in the table also shows that the calculated solubility is basically consistent with the results reported in the literature.

Table 4-1 Stability constant of metal ion hydroxyl complex(25℃)

Metal ion	$\lg K_1$	$\lg\beta_2$	$\lg\beta_3$	$\lg\beta_4$	pK_{sp}
Mg^{2+}	2.58	1.0			11.15
Ca^{2+}	1.4	2.77			5.22
Ba^{2+}	0.6				3.6
Mn^{2+}	3.4	5.8	7.2	7.3	12.6
Fe^{2+}	4.5	7.4	10.0	9.6	15.1
Co^{2+}	4.3	8.4	9.7	10.2	14.9
Ni^{2+}	4.1	8.0	11.0		15.2
Cu^{2+}	6.3	12.8	14.5	16.4	19.32
Zn^{2+}	5.0	11.1	13.6	14.8	15.52-16.46
Pb^{2+}	8.3	10.9	13.9		15.1-15.3
Cr^{3+}	9.99	11.88		29.87	30.27
Al^{3+}	9.01	18.7	27.0	33.0	33.5
Fe^{3+}	11.81	22.3	32.05	34.3	38.8
Ce^{3+}	5.9	11.7	16.0	18.0	21.9
Zr^{4+}	14.32	28.26	41.41	55.27	57.2
La^{3+}	5.5	10.8	12.1	19.1	22.3
Ti^{4+}	14.15	27.88	41.27	54.33	58.3

Table 4-2 Dissolved product of minerals and compounds

Compound	pK_{sp}	Compound	pK_{sp}	Compound	pK_{sp}
MnS(pink)	10.5	$MnCO_3$	9.3	$AlPO_4 \cdot 3H_2O$	18.24
MnS(green)	13.5	$ZnCO_3$	10.0	$Ca_{10}(PO_4)_6F_2$	118
FeS	18.1	$PbCO_3$	13.13	$Ca_{10}(PO_4)_6(OH)_2$	115
FeS_2	28.3	$CuCO_3$	9.63	$CaHPO_4$	7.0
CoS(α)	21.3	$CaCO_3$(calcite)	8.35	$FePO_4 \cdot 2H_2O$	36
CoS(β)	25.6	$CaCO_3$(nepheline)	8.22	Fe_2O_3	42.7
NiS(α)	19.4	$MgCO_3$	7.46	FeOOH	41.5
NiS(β)	24.9	$CoCO_3$	9.98	ZnO	16.66
NiS(γ)	26.6	$NiCO_3$	6.87	CaF_2	10.41
Cu_2S	48.5	$CaSO_4$	4.62	$ZnSiO_3$	21.03
CuS	36.1	$BaSO_4$	9.96	Fe_2SiO_4	18.92
$CuFeS_2$	61.5	$PbSO_4$	6.2	$CaSiO_3$	11.08
ZnS(α)	24.7	$SrSO_4$	6.5	$MnSiO_3$	13.20
ZnS(β)	22.5	$CaWO_4$	9.3	HgS	53.5
CdS	27.0	$MnWO_4$	8.85	Ag_2S	50.0
PbS	27.5	$FeWO_4$	11.04		

Table 4-3 Solubility of sulfide ore in pure water (mol/L)

Mineral	Chemical formula	Solubility	Mineral	Chemical formula	Solubility
Covellite	CuS	3.6×10^{-15}	Marcasite	FeS	3.6×10^{-6}
Chalcocite	Cu_2S	1.1×10^{-14}	Greenockite	CdS	1.23×10^{-10}
Chalcopyrite	$CuFeS_2$	1.9×10^{-14}	Chalcogenite	CoS	9.0×10^{-8}
Galena	PbS	7.9×10^{-11}	Nickel sulfide	NiS	8.1×10^{-7}
Sphalerite	ZnS	1.0×10^{-9}	Argentite	Ag_2S	1.4×10^{-17}
Pyrite	FeS_2	5.8×10^{-8}	Cinnabar	HgS	5.1×10^{-20}

4.1.2 Solubility of oxide minerals

The calculation of the solubility of oxide minerals mainly considers the hydrolysis reaction of metal ions. For example, the dissolution reaction of hematite in water is:

$$Fe_2O_3 + 3H_2O \rightleftharpoons 2Fe^{3+} + 6OH^- \qquad K_{sp} = 10^{-42.7} \tag{4-13}$$

$$\alpha_{Fe} = 1 + 10^{11.81}[OH^-] + 10^{22.3}[OH^-]^2 + 10^{32.05}[OH^-]^3 + 10^{34.4}[OH^-]^4$$

Since the concentration of OH^- produced by the dissolution of Fe_2O_3 is very small, the concentration of OH^- ions in the Fe_2O_3 saturated aqueous solution can be regarded as 1.0×10^{-7}, so that α_{Fe}=1.12×10^{11}, the solubility of Fe_2O_3 is $S_m = \left(K_{sp}\alpha_{Fe}/3^3\right)^{1/4} = 5.36\times10^{-9}$.

Another example is that the dissolution reaction has an effect on the pH value of the solution; the dissolution reaction of ZnO in water is as follows:

$$ZnO + H_2 \rightleftharpoons Zn^{2+} + 2OH^- \quad K_{sp} = 10^{-16.66} \tag{4-14}$$

Calculated approximate value of ZnO solubility $\overline{S}$:

$$\overline{S} = (K_{sp}/4)^{1/3} = 10^{-5.55} = 2.82 \times 10^{-6}$$

Thus $[OH^-] = 2\overline{S} = 5.64 \times 10^{-6}$, $pH = 8.75$. This is the approximate value of the OH^- ion concentration in the saturated aqueous solution of ZnO, and it is used to calculate the approximate value of α_{Zn}:

$$\alpha_{Zn} = 1 + 10^{5.0}[OH^-] + 10^{11.1}[OH^-]^2 + 10^{13.6}[OH^-]^3 + 10^{14.8}[OH^-]^4 = 5.54$$

To calculate the approximate value:

$$K'_{sp} = 10^{-16.66} \times 5.54 = 10^{-15.92}$$

$$S_m = K'_{sp}/[OH^-]^2 = 3.78 \times 10^{-6}$$

Taking $2S_m$ as the more accurate concentration of OH^-, α_{Zn}, K'_{sp} and S_m can be calculated sequentially. This method is called the step-by-step approximation method. Repeated calculation and repeated use can finally obtain a more accurate S_m value and the pH value of the saturated solution, which is recorded as pH_{sat}. In this example, the final S_m of ZnO is 3.5×10^{-6} mol/L, and pH_{sat} =8.85.

From the constants in Tables 4-1 and 4-2, according to the same method as above, the solubility and pH_{sat} value of the saturated solution of other oxide minerals can be calculated, see Table 4-4. It can be seen from the table that the solubility of oxide minerals is higher than that of sulfide, and the calculated values are basically consistent with the results reported in the literature.

$pH_{sat} = 7.00$ in the table means that the dissolution of the mineral basically has no effect on the pH of the solution.

Table 4-4 Solubility and pH_{sat} value of oxidized minerals

Oxidized minerals	Chemical formula	Solubility/(mol/L)	pH_{sat}
Hematite	Fe_2O_3	5.36×10^{-9}	~7.00
Gibbsite	$Al_2O_3 \cdot 3H_2O$	1.2×10^{-8}	~7.00
Zincite	ZnO	3.78×10^{-6}	8.85
Goethite	FeOOH	1.3×10^{-8}	~7.00
Cuprite	Cu_2O	2.24×10^{-7}	6.65
Valleriite	CuO	1.04×10^{-7}	7.01
Zircon	ZrO_2	1.2×10^{-12}	~7.00
Rutile	TiO_2	7.9×10^{-12}	~7.00
Cassiterite	SnO_2	2.22×10^{-13}	~7.00
Lead oxide	PbO	2.45×10^{-5}	9.69

4.1.3 Solubility of salt minerals

In the calculation of the solubility of salt minerals, the hydrolysis reaction of metal ions and the proton addition reaction of anions should be considered. Due to the large solubility product of such minerals, the protonation reaction of dissolved mineral anions generally has a large effect on pH.

(1) Carbonate minerals.

① In closed system. Take the solubility calculation of smithsonite as an example:

$$ZnCO_3 \rightleftharpoons Zn^{2+} + CO_3^{2-} \quad K_{sp} = 10^{-10.0} \tag{4-15}$$

The hydrolysis reaction of CO_3^{2-}:

$$CO_3^{2-} + H_2O \rightleftharpoons HCO_3^- + OH^- \tag{4-16}$$

$$\frac{[OH^-][HCO_3^-]}{[CO_3^{2-}][H_2O]} = \frac{K_W}{K_a} \tag{4-17}$$

Approximately considered that $[OH^-]=[HCO_3^-]$, then the initial concentration of $[CO_3^{2-}]$ can be approximately from $K_{sp,ZnCO_3}$, and the result is 10^{-5}. When balancing,

$$[CO_3^{2-}] = 10^{-5} - [OH^-] \tag{4-18a}$$

Put the above into equation (4-17):

$$[OH^-]^2 + 10^{-3.67}[OH^-] - 10^{-8.67} = 0$$

$$[OH^-] = 1.9\times10^{-5}\,mol/L \quad pH = 9.28$$

Get:

$$\alpha_{Zn} = 1 + 10^{5.0}\times1.9\times10^{-5} + 10^{11.1}\times(1.9\times10^{-5})^2 + 10^{13.6}\times(1.9\times10^{-5})^3 + 10^{14.8}\times(1.9\times10^{-5})^4 = 48.35$$

$$\alpha_{CO_3^{2-}} = 1 + 10^{10.33}\times10^{-9.28} + 10^{16.68}\times(10^{-9.28})^2 = 12.22$$

$$K'_{sp} = 10^{-10.0}\times48.35\times12.22 = 5.9\times10^{-8}$$

$$S_m = \sqrt{K'_{sp}} = 2.43\times10^{-4}\,mol/L$$

At $[CO_3^{2-}]' = S_m$, the score of $[CO_3^{2-}]$ is $1/\alpha_{CO_3^{2-}} = 1/12.22$. Thus, the concentration of free CO_3^{2-} is:

$$[CO_3^{2-}] = \frac{S_m}{12.22} = 1.99\times10^{-5}\,mol/L$$

Then bring in equation (4-17), and the more accurate $[OH^-]$ can be obtained as follows:

$$\frac{[OH^-]^2}{1.99\times10^{-5}-[OH^-]}=10^{-14.0}\times10^{10.33}=10^{-3.67}$$

$$[OH^-]=1.83\times10^{-5} \qquad pH=9.26$$

So get: $\alpha_{Zn}=45.0, \alpha_{CO_3^{2-}}=12.75$

$$S_m=\sqrt{10^{-10}\times45.0\times12.75}=2.4\times10^{-4}(mol/L)$$

Repeatedly using the step-by-step approximation method, finally the saturated solution of smithsonite is obtained, S_m is 2.4×10^{-4}, pH_{sat}=9.26.

② In open system. In an open system, since the dissolution of carbonate minerals is affected by the partial pressure of CO_2 in the atmosphere. There is the following formula:

$$[CO_3^{2-}][H^+]^2=10^{-21.64} \tag{4-18b}$$

The initial concentration of CO_3^{2-} is still approximately calculated from $K_{sp,ZnCO_3}$ as 10^{-5}, which is brought into Formula (4-18b) to obtain pH=8.32, thus:

$$\alpha_{Zn}=1+10^{5.0}\times10^{-5.68}+10^{11.1}\times(10^{-5.68})^2+10^{13.6}\times(10^{-5.68})^3+10^{14.8}\times(10^{-5.68})^4=1.76$$

$$\alpha_{CO_3^{2-}}=1+10^{10.33}\times10^{-8.32}+10^{16.68}\times(10^{-8.32})^2=104.4$$

$$K'_{sp}=10^{-10}\times1.76\times104.4=1.84\times10^{-8}$$

$$S_m=\sqrt{K'_{sp}}=1.3555\times10^{-4}mol/L$$

Then the more accurate concentration of free CO_3^{2-} ions is:

$$[CO_3^{2-}]=\frac{S_m}{104.4}=1.298\times10^{-6}$$

Bringing into formula (4-18b), a more accurate pH=7.88 can be obtained. In this way, the step-by-step approximation method is repeatedly used, and finally, in the open system, the S_m of smithsonite is 2.24×10^{-4}, and the pH value of the saturated solution is 7.65, which is more in line with reality. In Table 4-5, the pH value of carbonate mineral saturated solution is basically the same as the measured value based on the open system.

Table 4-5 Solubility and pH_{sat} value of salt minerals

Mineral	Chemical formula	Solubility/(mol/L)		pH_{sat}	
		Calculated value	Literature value	Close system	Open system
Magnesite	$MgCO_3$	3.22×10^{-4}		9.68	7.9
Fluoroapatite	$Ca_{10}(PO_4)_6F_2$	5.38×10^{-7}		7.4	
Hydroxyapatite	$Ca_{10}(PO_4)_6(OH)_2$	7.75×10^{-7}		7.6	
Fluorite	CaF_2	2.1×10^{-4}	2.0×10^{-4}	6.8	
Scheelite	$CaWO_4$	2.23×10^{-5}	$(1.74-3.5)\times10^{-5}$	6.42	
Plaster	$CaSO_4$	4.9×10^{-3}	7.8×10^{-3}	7.5	

(continued)

Mineral	Chemical formula	Solubility/(mol/L)		pH_{sat}	
		Calculated value	Literature value	Close system	Open system
Calcite	$CaCO_3$	1.18×10^{-4}	1.3×10^{-4}	9.73	8.2
Calcium silica	$CaSiO_3$	1.28×10^{-4}		10.4	
Barite	$BaSO_4$	1.05×10^{-5}	9.32×10^{-6}	9.19	
Celestite	$SrSO_4$	5.62×10^{-4}		8.29	
Malachite	$Cu_2CO_3(OH)_2$	4.5×10^{-7}		7.65	4.9
Cerussite	$PbCO_3$	8.29×10^{-6}	6.37×10^{-6}	7.43	6.65
Bonamite	$ZnCO_3$	2.24×10^{-4}	1.3×10^{-4}	9.26	7.65
Siderite	$FeCO_3$	2.00×10^{-5}	5.0×10^{-6}	9.23	7.44
Rhodochrosite	$MnCO_3$	5.23×10^{-5}		9.68	7.9
Spherocobaltite	$CoCO_3$	3.32×10^{-5}		9.57	7.7
Anglesite	$PbSO_4$	7.94×10^{-4}	1.45×10^{-4}	5.4	
Willemite	$ZnSiO_3$	1.55×10^{-7}		7.1	
Rhodonite	$MnSiO_3$	2.5×10^{-5}		9.4	
Fayalite	Fe_2SiO_4	2.5×10^{-6}			
Fischerite	$AlPO_4\cdot3H_2O$	2.5×10^{-7}		5.2	
Strengite	$FePO_4\cdot2H_2O$	1.6×10^{-5}			
Kaolin	$H_4Al_2Si_2O_9$	2.45×10^{-7}		5.6	

(2) Phosphate minerals. The dissolution equilibrium of fischerite is:

$$AlPO_4 \rightleftharpoons Al^{3+} + PO_4^{3-} \quad K_{sp} = 10^{-18.24} \tag{4-19}$$

$$PO_4^{3-} + H_2O \rightleftharpoons OH^- + HPO_4^{2-}$$

$$\frac{[OH^-][HPO_4^{2-}]}{[PO_4^{3-}][H_2O]} = \frac{K_W}{K_\alpha} = 10^{-14}/10^{-12.25} = 10^{-1.75} \tag{4-20}$$

The approximate value of the initial concentration of PO_4^{3-} calculated by K_{sp} is $10^{-9.12}$. During equilibrium:

$$[PO_4^{3-}] = 10^{-9.12} - [OH^-]$$

Then,

$$[OH^-]^2 + 10^{-1.65}[OH^-] - 10^{-10.77} = 0 \tag{4-21}$$

$$[OH^-] = 10^{-8.8}, pH_{sat} = 5.2$$

$$\alpha_{PO_4^{3-}} = 1 + 10^{12.35}[H^+] + 10^{10.55}[H^+]^2 + 10^{21.7}[H^+]^3 = 1.43\times10^9$$

$$\alpha_{Al} = 1 + 10^{9.9}[OH^-] + 10^{18.7}[OH^-]^2 + 10^{27.0}[OH^-]^3 + 10^{33.0}[OH^-]^4 = 19.2$$

$$K'_{sp} = 10^{-18.2}\times19.2\times1.43\times10^9 = 1.5\times10^{-8}$$

$$S_m = 1.26\times10^{-4}\,\mathrm{mol/L}$$

However, the exact calculation requires the use of the charge balance formula.

(3) Silicate minerals. The dissolution of silicate minerals is more complicated. A notable feature is that in an aqueous solution under certain conditions, the dissolution of complex silicate minerals proceeds step by step, sometimes producing another mineral. E.g:

The dissolution reaction of kaolin is carried out in two steps:

① $$H_4Al_2Si_2O_{9(S)} + 5H_2O \rightleftharpoons Al_2O_3\cdot3H_2O + 2H_4SiO_{4(aq)} \tag{4-22}$$

② in acid solution, $$Al_2O_3\cdot3H_2O + 6H^+ \rightleftharpoons 2Al^3 + 6H_2O \tag{4-23a}$$

In alkaline solution, $$Al_2O_3\cdot3H_2O + 2OH^- \rightleftharpoons 2Al(OH)_4^- \tag{4-23b}$$

The total reaction is:

In acid solution:

$$\frac{1}{2}H_4Al_2Si_2O_{9(S)} + 3H^+ = Al^{3+} + H_4SiO_4 + \frac{1}{2}H_2O, K_1 = 10^{3.35} \tag{4-24a}$$

In alkaline solution:

$$\frac{1}{2}H_4Al_2Si_2O_{9(S)} + \frac{5}{2}H_2O + OH^- = H_4SiO_4 + Al(OH)_4^-, K_2 = 10^{-5.7} \tag{4-24b}$$

The relationship between the solubility of kaolin and pH is:

$$\lg[Al]_T = \lg K_1 + \lg\alpha_{Al} - \lg[H_4Si_4] - 3pH \tag{4-25}$$

In pure water, assuming pH=7.0, then,

$$[Al]_T = 1.29\times10^{-7}\,\mathrm{mol/L}$$

(4) Sulfate minerals. Since sulfuric acid is a strong acid, the calculation of the solubility of such minerals only needs to consider the hydrolysis of cations. Take anglesite as an example:

$$PbSO_4 \rightleftharpoons Pb^{2+} + SO_4^{2-} \quad K_{sp} = 10^{-6.20} \tag{4-26}$$

$$Pb^{2+} + H_2O \rightleftharpoons PbOH^+ + H^+$$

$$\frac{[H^+][PbOH^+]}{[Pb^{2+}]} = K_W\cdot K_1 \tag{4-27}$$

Initial $[Pb^{2+}] = 10^{-3.1}$, and assuming $[H^+]=[PbOH^+]$, then

$$[H^+]+^2 + 10^{-7.7}[H^+] - 10^{-10.8} = 0 \tag{4-28}$$

Get: $[H^+] = 10^{-5.4}$, $pH = 5.4$

$$\alpha_{Pb} = 1 + 10^{6.3}[OH^-] + 10^{10.9}[OH^-]^2 + 10^{13.9}[OH^-]^3 = 1$$

So the solubility of $PbSO_{4(S)}$ is $S_m = \sqrt{K_{sp}} = 10^{-3.1} = 7.9\times10^{-4}(\mathrm{mol/L})$.

According to the method described above, from the data in Tables 4-1 and 4-2, the solubility of various salt minerals in pure water and the pH value of their saturated solution can be calculated, as shown in Table 4-5. It can be seen that the solubility of salt minerals is generally much greater than that of the corresponding sulfide and oxide minerals. Moreover, the pH of saturated solutions of salt minerals is either acidic or basic. This suggests that the flotation of salt minerals will be much more complicated than that of sulfide ores.

4.2 The Effect of Mineral Dissolution on Flotation

4.2.1 pH value of pulp and its buffering properties

It can be seen from Tables 4-1 to 4-3 that the dissolution of sulfide minerals generally has no effect on the pH value of the solution, and the dissolution of oxide minerals has little effect on the pH value of the solution, but the dissolution of most salt minerals increases or decreases the pH value of the solution. Since the pH value of the mineral saturated solution has a certain size, the pH value of the salt mineral pulp is generally maintained in a narrow range, which is the buffering property of the salt mineral pulp. This means that no matter what the initial pH value of the pulp is, after a certain period of equilibration, the pH value of the salt mineral pulp will eventually tend to a narrow range.

For example, the initial pH of the saturated solution of kaolin is 7.0. After about 1 minute, its pH drops to about 5.6, as shown in Figure 4-1, which is close to the calculated value in Table 4-5. Another typical example is calcite. As shown in Figure 4-2, the initial pH was adjusted to 8.2, 9.9 or 11.5 with NaOH, and the initial pH was adjusted to 3.0 and 1.9 with HNO_3. After several hours, the final equilibrium pH tended to be 8-9, which is close to the calculated value in Table 4-5.

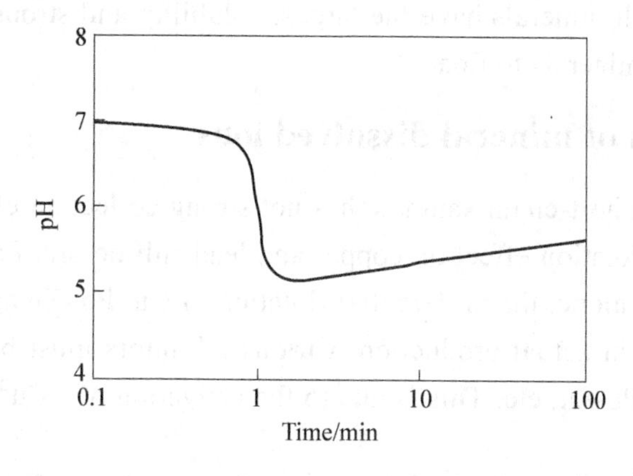

Figure 4-1 Change of pH value of kaolin saturated solution with contact time

These results show that the adjustment of pH in the flotation of salt minerals is complicated, and sometimes it is even difficult to adjust the pH value as we expect. As is the case with calcite, many experimental studies have shown that it is difficult to adjust the pH of calcite to below 7.0 during flotation.

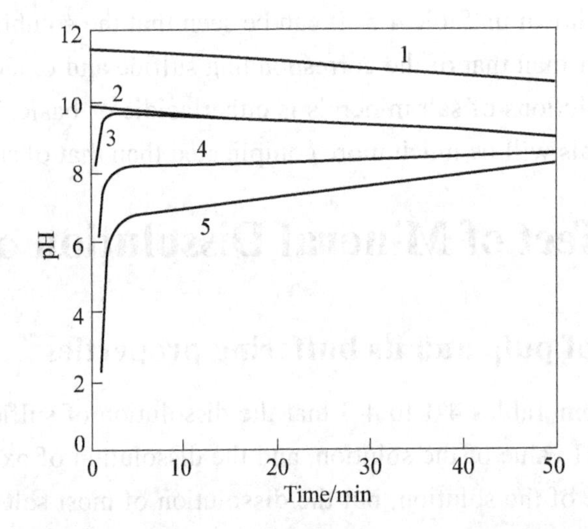

Figure 4-2 The relationship between the pH value of the calcite solution and the stirring time

Adjust initial pH with NaOH: 1—pH = 11.5; 2—pH = 9.9; 3—pH = 8.2

Adjust initial pH with HNO_3 : 4—pH = 3.0; 5—pH = 1.9

4.2.2 Mineral solubility and floatability

The solubility of minerals depends on the hydration of lattice ions and the interaction between ions, which are measured by hydration energy and lattice energy respectively. In general, minerals with high hydration energy have high solubility, high hydrophilicity and poor floatability. Therefore, the pure sulfide minerals which are difficult to dissolve show a certain natural floatability; the oxidized ore with high solubility is hydrophilic and has no natural floatability; salt minerals have the largest solubility and strong hydrophilicity. They are the most difficult minerals to float.

4.2.3 Activation of mineral dissolved ions

As we all know, short-chain xanthate has not strong collection effect on sphalerite and pyrite, but has good flotation effect on copper and lead sulfide ore. From the perspective of pure mineral flotation alone, the preferential flotation of Cu, Pb, Zn and Fe sulfide minerals is not a problem, but in actual production, various inhibitors must be added to realize the separation of Cu-Zn, Pb-Zn, etc. This is due to the activation of Cu^{2+}, Pb^{2+} and other ions dissolved in minerals.

The activation reaction of metal ions on sphalerite is carried out according to the following formula:

$$Me^{2+} + ZnS \rightleftharpoons MeS + Zn^{2+} \quad K_1 = \frac{[Zn^{2+}]}{[Me^{2+}]} \tag{4-29}$$

The standard free energy change of the reaction is:

$$\Delta G^{0'} = \Delta G^0_{Zn^{2+}} + \Delta G^0_{MeS} - \Delta G^0_{Me^{2+}} - \Delta G^0_{ZnS} \tag{4-30}$$

The ratio of metal ion concentration required for activation of sphalerite to Zn^{2+} concentration is:

$$\frac{[Me^{2+}]}{[Zn^{2+}]} = \exp\left(\frac{\Delta G^{0'}}{RT}\right) \tag{4-31}$$

Similarly, the reaction of pyrite activation is:

$$Me^{2+} + FeS_2 \rightleftharpoons MeS + Fe^{2+} + S^0$$

$$\Delta G^{0''} = \Delta G^0_{Fe^{2+}} + \Delta G^0_{MeS} - \Delta G^0_{FeS_2} - \Delta G^0_{Me^{2+}} \tag{4-32}$$

The ratio of the metal ion concentration required to activate pyrite to the Fe^{2+} ion concentration in the solution is:

$$[Me^{2+}]/[Fe^{2+}] = \exp\left(\frac{\Delta G^{0''}}{RT}\right) \tag{4-33}$$

According to the thermodynamic data, the ion free energies of Cu^{2+}, Pb^{2+},Cd^{2+}, Ag^+ are 15.33, −5.81, −18.58, and 18.43 kcal/mol, respectively. The ΔG^0 and required concentration of $Cu^{2+}, Pb^{2+}, Cd^{2+}, Ag^+$ ions for the activation reaction of sphalerite and pyrite are calculated from equations (4-31) and (4-33), as shown in Table 4-6.

Table 4-6 Conditions for activation of sphalerite and pyrite by metal ions

Metal ion	$-\Delta G^0$ /(kcal/mol)		Required for calculation		When two sulfides coexist	
	$-\Delta G^{0'}$	$-\Delta G^{0''}$	$\frac{[Me^{2+}]}{[Zn^{2+}]}$	$\frac{[Me^{2+}]}{[Fe^{2+}]}$	$\frac{[Me^{2+}]}{[Zn^{2+}]}$	$\frac{[Me^{2+}]}{[Fe^{2+}]}$
Cu^{2+}	15.04	11.53	9.3×10^{-12}	3.4×10^{-9}	10^{-5}	10^{-8}
Pb^{2+}	4.12	0.64	9.5×10^{-4}	0.34	7.9×10^{-2}	1.4×10^{-3}
Cd^{2+}	2.8	−0.68	8.8×10^{-3}	3.15	0.12	2.12×10^{-3}
Ag^+	34.26	29.06	7.4×10^{-26}	4.9×10^{-22}	10^{-8}	2.4×10^{-10}

It can be seen from Table 4-6 that the concentration of metal ions produced by the dissolution of sulfide ores such as Cu, Pb, Cd and Ag is much higher than that required for the activation of sphalerite. For pyrite, the activation of Cu^{2+} and Ag^+ ions is strong, while the activation of Pb^{2+} and Cd^{2+} ions is small.

In actual production, sulfide ore is easy to be oxidized; its solubility tends to increase. Table 4-7 lists the solubility of corresponding oxidation products of some sulfide ores. It can be seen from the table that the solubility of sulfide ore increases greatly after oxidation. For example, the solubility of $PbSO_4$ and $CuSO_4$ is 10^{7} and 10^{14} times higher than that of corresponding PbS and CuS. Therefore, after surface oxidation, the sulfide ore loses its natural floatability, but it is conducive to its floating because the sulfide ore is easy to interact with the collector.

Table 4-7 The solubility of several typical sulfide minerals and their sulfates

Sulfide minerals	Solubility/(mol/L)	Sulfate	Solubility/(mol/L)	Ratio of solubility of two substances
CuS	3.6×10^{-15}	$CuSO_4$	1.08（78℃）	3.0×10^{14}
PbS	7.9×10^{-11}	$PbSO_4$	7.94×10^{-4}（25℃）	10^7
ZnS	1.0×10^{-9}	$ZnSO_4$	3.3（18℃）	3.3×10^9
FeS_2	5.8×10^{-8}	$FeSO_4$	1.031	1.8×10^7
FeS	3.6×10^{-6}			2.9×10^5

On the other hand, after the sulfide ore is oxidized, the ion concentration in the pulp increases, and the mutual influence is strengthened.

It is reported that in the presence of oxygen, in the saturated solution of chalcopyrite, the coverage of Cu on the surface of sphalerite can reach 1.5 monolayers, and copper blue is formed, as shown in Figure 4-3.

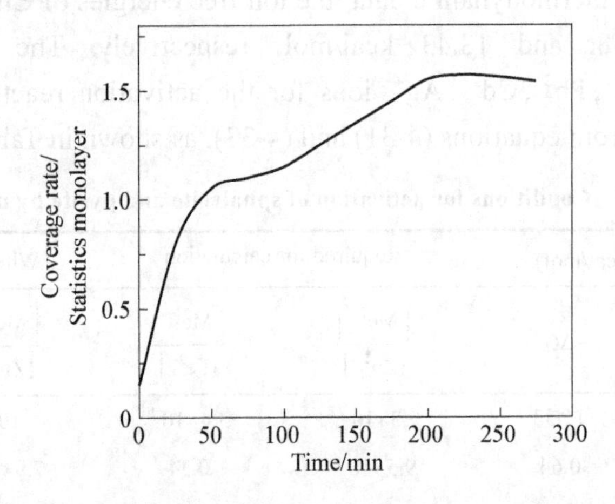

Figure 4-3 Copper coverage of sphalerite surface in chalcopyrite saturated solution

Since most sulfide ores are polymetallic ores, i.e. sulfide ores such as copper, lead, zinc and iron are often symbiotic, and some ores also contain Au, Ag, etc. Therefore, when using xanthate collectors for flotation, various inhibitors, such as NaCN, lime, $ZnSO_4$, etc., must be added for the separation of polymetallic sulfide ores. In addition to directly inhibiting sphalerite and pyrite, these reagents can also effectively prevent the activation of other ions in the pulp.

For example, in Cu-Zn flotation separation, NaCN is used as an depressants, and the control reaction is usually as follows:

$$Zn^{2+} + 4CN^{-} = Zn(CN)_4^{2-}$$

The prevention reaction of activation is:

$$Cu^{2+} + 4CN^- = Cu(CN)_4^{2-}$$

For carbonate, phosphate, sulfate, silicate, tungstate and other salt minerals, their solubility is large. It can be seen from Table 4-5 that they are generally 10^{-6} - 10^{-3}mol / L, which is within the concentration range of activated quartz, feldspar, beryl and other minerals for flotation. Silicate minerals such as quartz are also common gangue minerals. Therefore, in actual production, in order to achieve the separation of useful minerals from silicate gangues such as quartz, depressants must be added, such as sodium silicate and sodium hexametaphosphate, which also have the effects of depression and deactivation.

4.2.4 Influence of mineral dissolved ions on collector action

(1) Competitive adsorption. The mineral dissolved ions and the collector ions of the same electricity compete for adsorption on the mineral surface, which inhibits the flotation. For example, using potassium oleate (KOl) as collector, the recovery rate of apatite flotation in apatite and calcite saturated solution is lower than that in pure water (Fig. 4-4).

It is believed that this is due to the competitive adsorption of PO_4^{3-} and CO_3^{2-} ions with oleate anion in the saturated solution of the two minerals. According to the data in Table 4-5, in the saturated solution of calcite and apatite, the concentration of CO_3^{2-}, PO_4^{3-} ions can reach 10^{-6} - 10^{-4}mol / L, which is greater than or close to the concentration of KOl in Figure 4-4 which is enough to produce competitive effect.

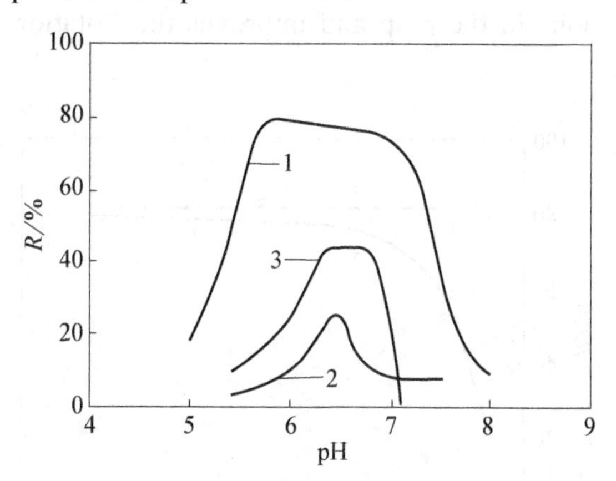

Figure 4-4 Relationship between flotation recovery of apatite and pH value

[KOl] 2×10^{-6}mol/L, [KNO_3] 3×10^{-6}mol/L;

1—in pure water; 2—calcite in saturated solution; 3—apatite saturated solution

Figure 4-5 shows the results of flotation of fluorite with dodecylamine. The flotation recovery in saturated solution of fluorite is lower than that in pure water. It can also be considered that it is due to the competition between Ca^{2+} and RNH_3^+. Because the concentration of Ca^{2+} ion in fluorite saturated solution can reach 10^{-4} mol/L, when the concentration of dodecylamine is greater than 10^{-4} mol/L, the flotation results in the two solutions are basically the same.

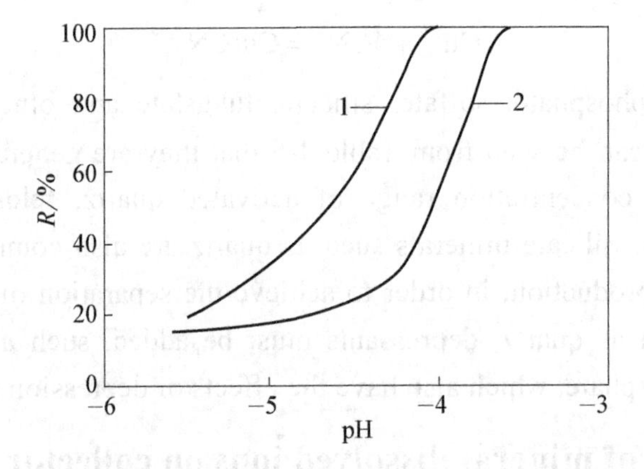

Figure 4-5 Relationship between flotation recovery of fluorite and pH value

1—in pure water; 2—fluorite in saturated solution

(2) Precipitation collector. Dissolved mineral cations can form salt precipitation with collector anions in solution, which consumes collector. It has an adverse impact on flotation, and increases the dosage. For example, the flotation of apatite/calcite with sodium oleate has low recovery. Others believe that it is not due to the competitive adsorption of CO_3^{2-} and Ol^-, but due to the formation of $Ca(Ol)_2$ precipitation between dissolved Ca^{2+} and Ol^-. Therefore, when a cation exchanger zeolite is added, the excess Ca^{2+} in the solution is exchanged with the Na^+ in the zeolite, which greatly reduces the concentration of Ca^{2+} ions in the pulp and improves the flotation recovery of apatite, as shown in Figure 4-6.

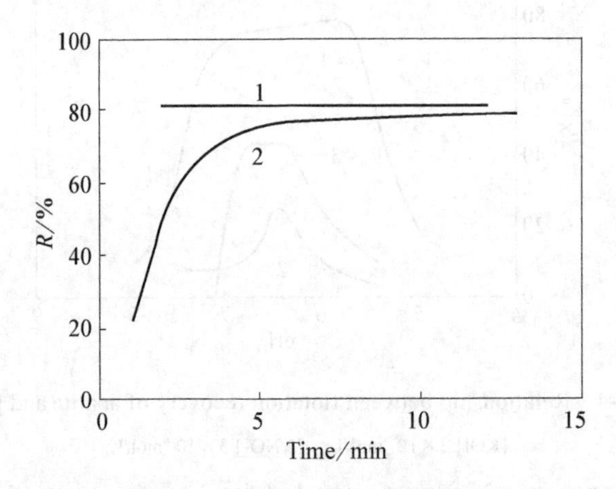

Figure 4-6 The flotation of apatite/calcite with sodium oleate

1—with 0.12g/L addition zeolite; 2—without zeolite

(3) Effect on the adsorption of collector. As can be seen from Figure 4-7, at pH between 4.5 and 5.5, the adsorption capacity of sodium dodecylbenzene sulfonate (DDBS) on kaolin decreased rapidly. The solubility curve of kaolin is drawn from formula (4-25), which is also shown in Figure 4-7. The solubility curve has a relatively

consistent relationship with the adsorption curve. Dissolved Al ions can affect the adsorption of collector from several aspects. It can absorb kaolin surface and increase its positive potential. It can also react with DDBS on the surface area, all of which are beneficial to the adsorption of DDBS. However, it can also form precipitation with DDBS in solution. Figure 4-7 shows that the adsorption capacity of DDBS decreases with the decrease of the solubility of kaolin. Obviously, the first two functions are dominant in the system.

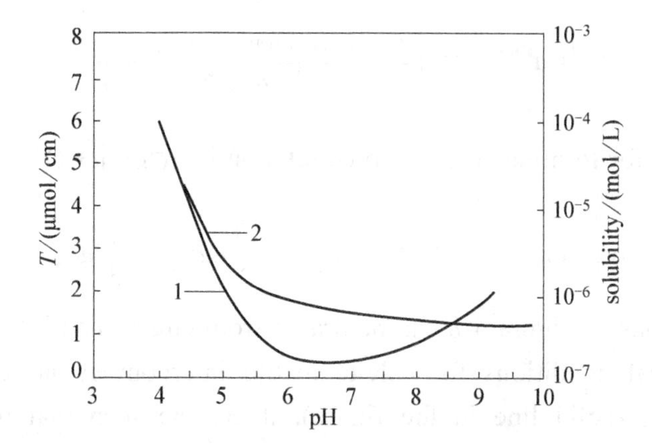

Figure 4-7 Relationship between the solubility of kaolin and the adsorption capacity of sodium dodecylbenzene sulfonate on kaolin and pH (25℃, S/L=0.2, [NaCl] =10^{-1}mol/L)

1—adsorption capacity; 2—solubility

4.3 Chemical Reactions and Surface Mutual Transformation of Mineral Dissolved Ions in Mineral Surfaces

The dissolved ions of one mineral are adsorbed in the surface of another mineral, and then a chemical reaction occurs, which will change the surface properties of the latter mineral, and even become similar to the surface of the former mineral.

For example, the following equilibrium exists in a saturated solution of calcite and apatite:

$$10CaCO_{3(S)} + 6PO_4^{3-} + 2OH^- \rightleftharpoons Ca_{10}(PO_4)_6(OH)_{2(S)} + 10CO_3^{2-} \tag{4-34}$$

$$K = \frac{[CO_3^{2-}]^{10}}{[PO_4^{3-}]^6[OH^-]^2} = \frac{K_{spCaCO_3}}{K_{spCa_{10}(PO_4)_6(OH)_2}} = \frac{10^{-33.5}}{10^{-115}} = 10^{81.5} \tag{4-35}$$

From equation (4-18b), get:

$$\lg\left[CO_3^{2-}\right] = -21.64 + 2pH$$

Therefore, in atmosphere, the critical condition for the conversion of calcite surface to apatite due to PO_4^{3-} produced by the dissolution of apatite is:

$$\lg\left[PO_4^{3-}\right]_T = \frac{5}{3}\lg\left[CO_3^{2-}\right] - \frac{1}{3}\lg\left[OH^-\right] - \frac{31.5}{6} = -36.65 + 3pH \tag{4-36}$$

The critical conditions for the conversion of apatite to calcite can be discussed as follows:

The concentration of Ca^{2+} ion produced by apatite dissolution is:

$$\left[Ca^{2+}\right] = \frac{10}{\alpha_{C_a}}\left[\frac{K_{spCa_{10}(PO_4)_6(OH)_2}\alpha_{Ca^{2+}}^{10}\alpha_{PO_4^{3-}}^{6}}{10^{10}\times 6^6 \times 2^2}\right]^{1/2} \tag{4-37}$$

The conditions for forming $CaCO_3$ precipitation is $\left[Ca^{2+}\right]\left[CO_3^{2-}\right]_T = K_{spCaCO_3}$, then,

$$\lg\left[CO_3^{2-}\right]_T = -2.11 + \lg a_{CO_3^{2-}} + \frac{13}{9}\lg a_{Ca^{2+}} - \frac{1}{3}\lg a_{PO_4^{3-}} \tag{4-38}$$

Two dashed lines in Figure 4-8 can be drawn from equations (4-36) and (4-38), which represent the critical conditions for calcite-apatite interconversion. Compared with the experimental results (solid line in the figure), it can be seen that under a certain pH condition, when the concentration of dissolved CO_3^{2-} is large enough, the surface of apatite can become calcite. Similarly, when the dissolved PO_4^{3-} concentration is large enough, the calcite surface can become apatite. As shown in Figure 4-9, the zeta potential of apatite in calcite saturated solution is much higher than that in pure water, and in alkaline medium, its zeta potential curve is similar to that of calcite surface. The mutual transformation of such mineral surfaces often makes flotation separation more difficult. This problem is more serious because of the high solubility of salt minerals.

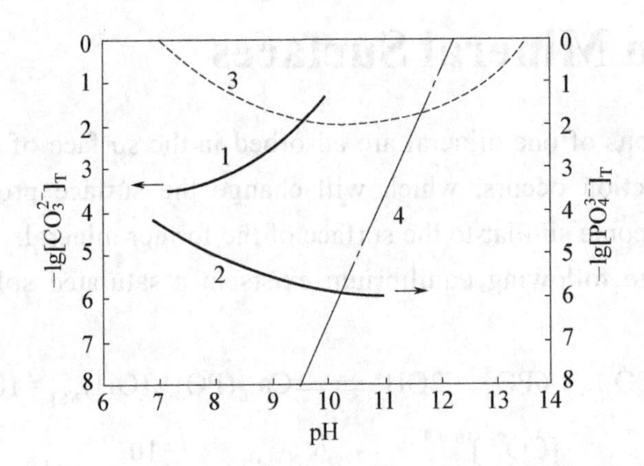

Figure 4-8 Conversion conditions of apatite and calcite

(the dotted line is the calculated conversion condition)

1—calcite in water; 2—apatite in water; 3—conversion of apatite to calcite;

4—conversion of calcite to apatite

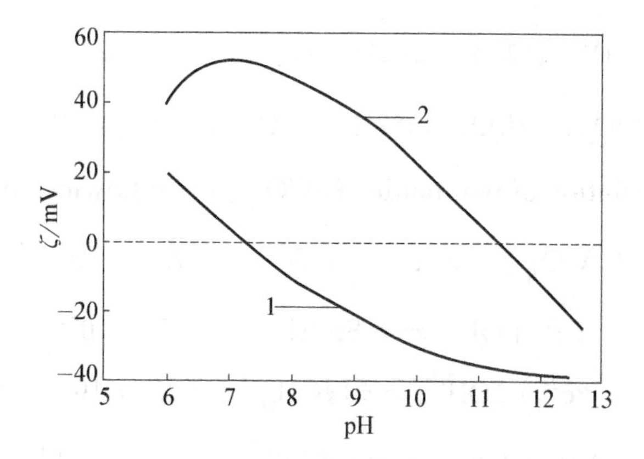

Figure 4-9 Relationship between zeta potential and pH of apatite

1—in pure water; 2—calcite in saturated solution

4.4 Logarithmic Concentration Diagram of Mineral Dissolved Components

In addition to calculating the total solubility of minerals, solution equilibrium can also calculate the concentration of each dissolved component, and draw $\lg C$-pH diagram. Taking tungstate minerals as an example, the $\lg C$-pH diagram of mineral dissolved components and its flotation significance are discussed below.

Tungstate minerals mainly refer to wolframite $(Mn,Fe)WO_4$ and scheelite $CaWO_4$. When discussing their dissolution equilibrium, it is explained by the dissolution equilibrium of $MnWO_4$, $FeWO_4$ and $CaWO_4$.

In the saturated solution of wolframite $MnWO_4$, the following equilibrium exists:

$$MnWO_{4(S)} \rightleftharpoons Mn^{2+} + WO_4^{2-} \qquad K_{sp1} = 10^{-8.85} \tag{4-39}$$

$$Mn^{2+} + OH^- \rightleftharpoons MnOH^+ \qquad K_1 = 10^{8.54} \tag{4-40}$$

$$Mn^{2+} + 2OH^- \rightleftharpoons Mn(OH)_{2(aq)} \qquad K_2 = 10^{5.8} \tag{4-41}$$

$$Mn^{2+} + 3OH^- \rightleftharpoons Mn(OH)_3^- \qquad K_3 = 10^{7.2} \tag{4-42}$$

$$Mn^{2+} + 4OH^- \rightleftharpoons Mn(OH)_4^{2-} \qquad K_4 = 10^{7.3} \tag{4-43}$$

$$2Mn^{2+} + OH^- \rightleftharpoons Mn_2(OH)^{3+} \qquad K_5 = 10^{4.13} \tag{4-44}$$

$$2Mn^{2+} + 3OH^- \rightleftharpoons Mn_2(OH)_3^+ \qquad K_6 = 10^{16.58} \tag{4-45}$$

$$Mn(OH)_{2(S)} \rightleftharpoons Mn^{2+} + 2OH^- \qquad K_{s_1} = 10^{12.6} \tag{4-46}$$

$$H^+ + WO_4^{2-} \rightleftharpoons HWO_4^- \qquad K_1^H = 10^{8.5} \tag{4-47}$$

$$H^+ + HWO_4^- \rightleftharpoons H_2WO_{4(aq)} \qquad K_2^H = 10^{4.6} \tag{4-48}$$

$$WO_{3(S)} + H_2O \rightleftharpoons 2H^+ + WO_4^{2-} \qquad K_{SO} = 10^{-14.05} \tag{4-49}$$

In the saturated solution of wolframite $FeWO_{4(S)}$ the following equilibrium exists:

$$FeWO_{4(S)} \rightleftharpoons Fe^{2+} + WO_4^{2-} \qquad K_{sp2} = 10^{-11.04} \tag{4-50}$$

$$Fe^{2+} + OH^- \rightleftharpoons FeOH^+ \qquad K_1' = 10^{4.5} \tag{4-51}$$

$$Fe^{2+} + 2OH^- \rightleftharpoons FeOH_{2(aq)} \qquad K_2' = 10^{7.2} \tag{4-52}$$

$$Fe^{2+} + 3OH^- \rightleftharpoons Fe(OH)_3^- \qquad K_3' = 10^{11.0} \tag{4-53}$$

$$Fe^{2+} + 4OH^- \rightleftharpoons Fe(OH)_4^{2-} \qquad K_4' = 10^{10.0} \tag{4-54}$$

$$Fe(OH)_{2(S)} \rightleftharpoons Fe^{2+} + 2OH^- \qquad K_{S_2} = 10^{-14.95} \tag{4-55}$$

It also includes the balance of equations (4-47) to (4-49).

In the saturated solution of $CaWO_{4(S)}$ the following equilibrium exists:

$$CaWO_{4(S)} \rightleftharpoons Ca^{2+} + WO_4^{2-} \qquad K_{sp3} = 10^{-9.3} \tag{4-56}$$

$$Ca^{2+} + OH^- \rightleftharpoons CaOH^+ \qquad K_1'' = 10^{1.40} \tag{4-57}$$

$$Ca^{2+} + 2OH^- \rightleftharpoons Ca(OH)_{2(aq)} \qquad K_2'' = 10^{2.77} \tag{4-58}$$

$$Ca(OH)_{2(S)} \rightleftharpoons Ca^{2+} + 2OH^- \qquad K_{S_3} = 10^{-5.22} \tag{4-59}$$

It also includes the balance of equations (4-47) to (4-49). From these equilibrium relations, it can be seen that in the saturated solution of $MnWO_4$, $FeWO_4$ and $CaWO_4$, tungstic acid and hydroxide will be precipitated under certain conditions. Therefore, their dissolution equilibrium must be considered separately.

(1) Formation of metal hydroxide precipitates. The pH at which $MnWO_4$ forms $Mn(OH)_2$ precipitation is determined by the following formula:

$$MnWO_{4(S)} + 2OH^- \rightleftharpoons Mn(OH)_{2(S)} + WO_4^{2-} \qquad K_{11} = 10^{8.75} \tag{4-60}$$

$$\frac{[WO_4^{2-}]}{[OH^-]^2} = 10^{8.75}, \frac{\sqrt{K_{sp1}}}{[OH^-]^2} = 10^{8.75}$$

get $pH_S = 9.9$.

When pH>9.9, $MnWO_4$ becomes $Mn(OH)_2$, and each component is in equilibrium with $Mn(OH)_2$ solid. The concentration of each component is:

$$\lg[Mn^{2+}] = 15.4 - 2pH \tag{4-61}$$

$$\lg[WO_4^{2-}] = -24.25 + 2pH \tag{4-62}$$

$$\lg[MnOH^+] = 4.94 - pH \tag{4-63}$$

$$\lg\left[Mn(OH)_{2(aq)}\right]=-6.8 \tag{4-64}$$

$$\lg\left[Mn(OH)_3^-\right]=-19.4+pH \tag{4-65}$$

$$\lg\left[Mn(OH)_4^{2-}\right]=-32.5+2pH \tag{4-66}$$

$$\lg\left[Mn_2(OH)_3^+\right]=5.33-pH \tag{4-67}$$

$$\lg\left[HWO_4^-\right]=-20.75+pH \tag{4-68}$$

The pH at which $FeWO_{4(S)}$ forms $Fe(OH)_{2(S)}$ precipitation is determined by the following equation:

$$FeWO_{4(S)}+2OH^- \rightleftharpoons Fe(OH)_{2(S)}+WO_4^{2-} \qquad K_{12}=10^{3.91} \tag{4-69}$$

$$\frac{\left[WO_4^{2-}\right]}{[OH^-]^2}=10^{3.91}, \frac{\sqrt{K_{sp2}}}{[OH^-]^2}=10^{3.91}$$

get pH_S=9.3.

Then when pH > 9.3, in $FeWO_{4(S)}$ saturated solution, each component is in equilibrium with $Fe(OH)_{2(S)}$ from Equations (4-47) to (4-49) and (4-50) to (4-55) find out the relationship between the concentration of each component and pH as:

$$\lg\left[HWO_4^-\right]=-20.75+pH \tag{4-70}$$

$$\lg[Fe^{2+}]=13.05-2pH \tag{4-71}$$

$$\lg\left[WO_4^{2-}\right]=-24.09+2pH \tag{4-72}$$

$$\lg\left[HWO_4^-\right]=-20.59+pH \tag{4-73}$$

$$\lg[FeOH^+]=3.55-pH \tag{4-74}$$

$$\lg\left[Fe(OH)_{2(aq)}\right]=-7.75 \tag{4-75}$$

$$\lg\left[Fe(OH)_3^-\right]=-17.95+pH \tag{4-76}$$

$$\lg\left[Fe(OH)_4^{2-}\right]=-32.95+2pH \tag{4-77}$$

(2) Conditions for the formation of $H_2WO_{4(S)}$ precipitates. The pH at which $MnWO_{4(S)}$ forms $H_2WO_{4(S)}$ precipitation is determined by the following formula:

$$MnWO_{4(S)}+2H^+ \rightleftharpoons H_2WO_{4(S)}+Mn^{2+} \qquad K_{13}=10^{5.2} \tag{4-78}$$

$$\frac{[Mn^{2+}]}{[H^+]^2}=10^{5.2}, \frac{\sqrt{K_{sp1}}}{[H^+]^2}=10^{5.2}$$

get $pH_m=4.8$.

When pH<4.8, the concentration of each component is determined by the dissolution balance of $H_2WO_{4(s)}$:

$$\lg\left[WO_4^{2-}\right]=-14.05+2pH \tag{4-79}$$

$$\lg\left[HWO_4^-\right]=-10.55+pH \tag{4-80}$$

$$\lg\left[H_2WO_4\right]=-5.95 \tag{4-81}$$

$$\lg\left[Mn^{2+}\right]=5.2-2pH \tag{4-82}$$

$$\lg\left[MnOH^+\right]=-5.26-pH \tag{4-83}$$

$$\lg\left[Mn_2OH^{3+}\right]=0.53-3pH \tag{4-84}$$

The pH at which $FeWO_{4(s)}$ forms $H_2WO_{4(s)}$ precipitation is determined by the following formula:

$$FeWO_{4(s)}+2H^+ \rightleftharpoons H_2WO_{4(s)}+Fe^{2+} \quad K_{14}=10^{3.01} \tag{4-85}$$

get $pH_m=4.3$.

When pH<4.3, the concentration of each component in $FeWO_{4(s)}$ saturated solution is also determined by the dissolution equilibrium equation (4-49) of $H_2WO_{4(s)}$:

$$\lg\left[Fe^{2+}\right]=3.01-2pH \tag{4-86}$$

$$\lg\left[FeOH^+\right]=-6.49-pH \tag{4-87}$$

(3) No formation of hydroxide and H_2WO_4 precipitates. In $MnWO_4$ saturated solution with 4.8 < pH < 9.9, there is no $H_2WO_{4(s)}$ and $Mn(OH)_{2(s)}$ is formed, at this time:

$$\left[Mn(OH)_{2(s)}\right]=0,\left[H_2WO_{4(s)}\right]=0$$

At this time, the total concentration of Mn (Ⅱ) component is calculated from the conditional solubility product of $Mn(OH)_{4(s)}$. The conditional solubility product of $Mn(OH)_{4(s)}$ is:

$$K'_{sp}=K_{sp1}\bullet\alpha_{Mn^{2+}}\bullet\alpha_{WO_4^{2-}} \tag{4-88}$$

Let $[Mn^{2+}]'$ be the total concentration of Mn (Ⅱ) component, then

$$[Mn^{2+}]'=\sqrt{K'_{sp1}}=\sqrt{K_{sp1}\alpha_{Mn^{2+}}\alpha_{WO_4^{2-}}} \tag{4-89}$$

According to the definition of the side reaction coefficient, have:

$$[Mn^{2+}]=[Mn^{2+}]'/\alpha_{Mn^{2+}}=\sqrt{K_{sp1}\frac{\alpha_{WO_4^{2-}}}{\alpha_{Mn^{2+}}}} \tag{4-90}$$

Therefore, the concentration of each component is:

$$\lg[Mn^{2+}] = \frac{1}{2}\left[\lg K_{sp1} + \lg\alpha_{WO_4^{2-}} - \lg\alpha_{Mn^{2+}}\right] \tag{4-91}$$

$$\lg\left[WO_4^{2-}\right] = \frac{1}{2}\left[\lg K_{sp1} + \lg\alpha_{Mn^{2+}} - \lg\alpha_{WO_4^{2-}}\right] \tag{4-92}$$

$$\lg[MnOH^{+}] = -10.46 + pH + \lg[Mn^{2+}] \tag{4-93}$$

$$\lg\left[Mn(OH)_{2(aq)}\right] = -22.2 + 2pH + \lg[Mn^{2+}] \tag{4-94}$$

$$\lg\left[Mn_2(OH)_3^{+}\right] = -25.47 + 3pH + 2\lg[Mn^{2+}] \tag{4-95}$$

$$\lg\left[HWO_4^{-}\right] = 3.5 - pH + \lg\left[WO_4^{2-}\right] \tag{4-96}$$

$$\lg\left[H_2WO_{4(aq)}\right] = 8.1 - 2pH + \lg\left[WO_4^{2-}\right] \tag{4-97}$$

$$\lg\left[Mn(OH)_3^{-}\right] = -34.8 + 3pH + \lg[Mn^{2+}] \tag{4-98}$$

In the same way, the concentration of each component in $FeWO_{4(S)}$ saturated solution at 4.3 < pH < 9.3:

$$\lg[Fe^{2+}] = \frac{1}{2}\left[\lg K_{sp2} + \lg\alpha_{WO_4^{2-}} - \lg\alpha_{Fe^{2+}}\right] \tag{4-99}$$

$$\lg\left[WO_4^{2-}\right] = \frac{1}{2}\left[\lg K_{sp2} + \lg\alpha_{Fe^{2+}} - \lg\alpha_{WO_4^{2-}}\right] \tag{4-100}$$

$$\lg[FeOH^{+}] = -9.5 + pH + \lg[Fe^{2+}] \tag{4-101}$$

$$\lg\left[Fe(OH)_{2(aq)}\right] = -20.8 + 2pH + \lg[Fe^{2+}] \tag{4-102}$$

$$\lg\left[Fe(OH)_3^{-}\right] = -31 + 3pH + \lg[Fe^{2+}] \tag{4-103}$$

$$\lg\left[Fe(OH)_4^{2-}\right] = -46 + 4pH + \lg[Fe^{2+}] \tag{4-104}$$

$$\lg\left[HWO_4^{-}\right] = 3.5 - pH + \lg\left[WO_4^{2-}\right] \tag{4-105}$$

$$\lg[H_2WO_4] = 8.1 - 2pH + \lg\left[WO_4^{2-}\right] \tag{4-106}$$

So far, the relationship between the concentration of each component and pH value in the three regions in the saturated solution of $MnWO_4$ and $FeWO_4$ is calculated. And draw Figures 4-10 and 4-11, i.e. lg$C$-pH diagram of each component in $MnWO_4$ and $FeWO_4$ saturated solution.

Using Figures 4-10 and 4-11, the change of zeta potential on the surface of huebnerite and ferberite with pH value can be theoretically analyzed.

① For huebnerite.

i. pH < 2.8, Mn^{2+} will dissolve from the surface in large quantities, and the potential determining ions on the surface should be $MnOH^{+}, HWO_4^{-}$. Under this condition,

$[MnOH^+]>[HWO_4^-]$, the surface of wolframite should be positively charged; when $[MnOH^+]=[HWO_4^-]$, pH = 2.8, which is the theoretical isoelectric point.

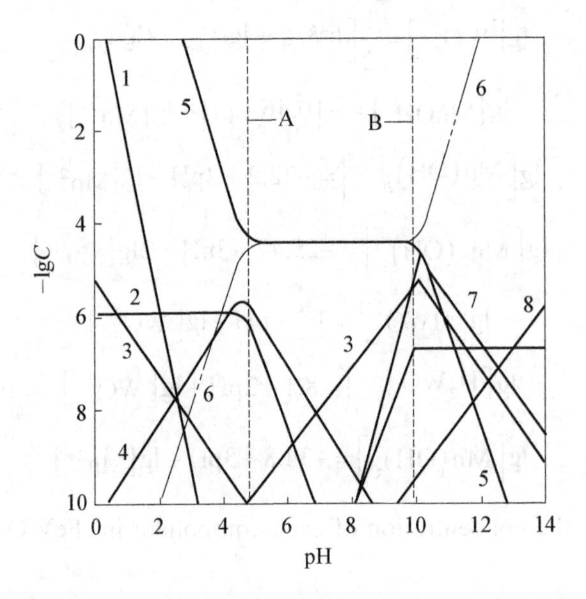

Figure 4-10 The logarithmic concentration diagram of $MnWO_{4(s)}$

A—pH_m: $MnWO_{4(s)} \leftrightarrow WO_{3(s)}$; B—pH_s: $MnWO_{4(s)} \leftrightarrow Mn(OH)_{2(s)}$

1— Mn_2OH^{3+} ; 2— $H_2WO_{4(s)}$; 3— $MnOH^+$; 4— HWO_4^- ; 5— Mn^{2+} ; 6— WO_4^{2-} , 7— $Mn_2(OH)_3^+$; 8— $Mn(OH)_3^-$

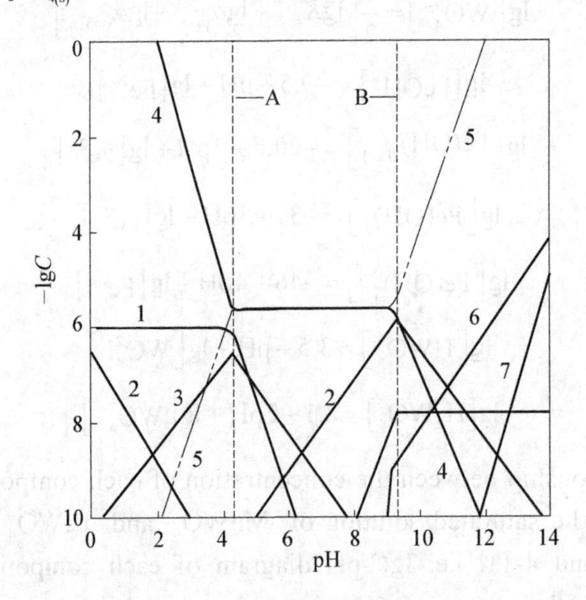

Figure 4-11 The logarithmic concentration diagram of $FeWO_{4(s)}$

A—pH_m: $MnWO_{4(s)} \leftrightarrow WO_{3(s)}$; B—pH_s: $MnWO_{4(s)} \leftrightarrow Mn(OH)_{2(s)}$

1—$H_2WO_{4(aq)}$; 2—$FeOH^+$; 3— HWO_4^- ; 4—Fe^{2+};5— WO_4^{2-} ; 6— $Fe(OH)_3^-$; 7— $Fe(OH)_4^{2-}$

ii. 2.8<pH<4.8, $[HWO_4^-]$ and $[WO_4^{2-}]$ are greater than $[MnOH^+]$, the negative value of zeta-potential in the surface of wolframite will increase. Until pH=4.8, $[HWO_4^-]$

has a maximum value, correspondingly, the negative zeta-potential also has a maximum value at this time.

iii. 4.8<pH<6, the concentration of $\left[HWO_4^-\right]$ decreases, the concentration of $\left[MnOH^+\right]$ increases, and the negative value of ζ-potential should decrease.

iv. In the range of pH=6.0-9.9, the dissolution of Mn^{2+} from the surface is greatly reduced, and $\left[Mn^{2+}\right]=\left[WO_4^{2-}\right]\gg\left[HWO_4^-\right]$ and $\left[MnOH^+\right]$. At this time, the ζ-potential of the mineral surface is determined by Mn^{2+} and WO_4^{2-}. Since Mn^{2+} is easy to dissolve from the surface relative to WO_4^{2-} ions, the surface potential of wolframite still has a small negative value in this pH range, which is the so-called "near PZC region".

v. pH>9.9=pH_S, the concentration of WO_4^{2-} increases sharply, and the negative value of ζ-potential on the surface of $MnWO_{4(S)}$ also increases sharply. At this time, the potential determining ion is WO_4^{2-} ion.

② For ferberite.

i. pH < 2.0, the surface is positively charged, and the potential determining ions are $FeOH^+$ and HWO_4^-. The IEP is $\left[FeOH^+\right]=\left[HWO_4^-\right]$. pH = 2.0.

ii. pH>2.0, the concentration of $\left[HWO_4^-\right]$ is greater than the concentration of $FeOH^+$, the surface of ferberite is negatively charged, and at pH=4.3, the HWO_4^- concentration and the negative value of zeta-potential are both maximum values.

iii. 6>pH>4.3, the zeta-potential decreases with the concentration of $\left[HWO_4^-\right]$.

iv. The pH of the "near PZC region" is 6-9.3.

v. pH>9.3, the negative value of zeta-potential on the surface of ferberite increases sharply with the concentration of WO_4^{2-}.

The relationship between the zeta potential of wolframite and pH measured experimentally is shown in Figure 4-12. It can be seen that the change of zeta potential of natural huebnerite with pH value is consistent with the result of solution equilibrium analysis. IEP=2.8, the maximum negative value of zeta potential appears at pH_m=4.8, the "near PZC region" is pH=6.0-9.8, and the negative value of zeta potential rises sharply at pH=9.8. Another natural wolframite (Mn:Fe=1.2) has IEP=2.6, which is between the theoretical IEP of $MnWO_{4(S)}$ and $FeWO_{4(S)}$, and the pH_m=4.6, the near PZC region is 6.0-9.5, and the pH=9.5 where the negative value of ζ-potential rises sharply is between the theoretical prediction values of wolframite and huebnerite. Therefore, it can be considered that with the increase of Fe content in wolframite, its electrical properties will be closer to $FeWO_{4(S)}$, and with the increase of manganese content, its electrical properties will be closer to $MnWO_{4(S)}$. Also, the pH at which the "near zero point region" occurs will always be around neutral pH.

The results of flotation of wolframite with several anionic collectors are shown in Figure 4-13. It can be noted that the optimum flotation pH range of wolframite is in the "near zero electric point region". The following is discussed according to Figures 4-10 to 4-12.

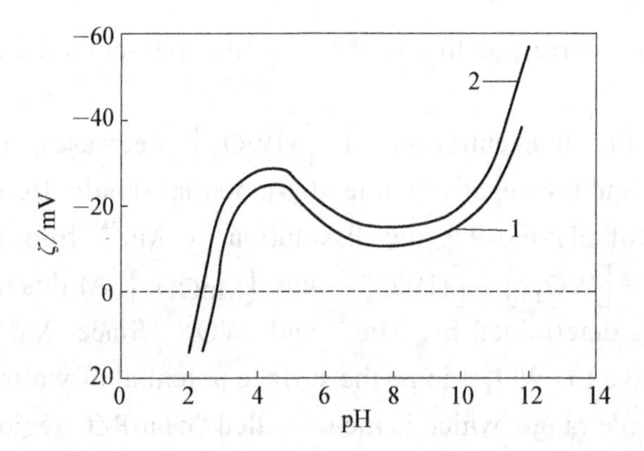

Figure 4-12 Relationship between zeta potential of wolframite, huebnerite and pH
1—wolframite, $Mn/Fe = 1.2$; 2—huebnerite ($FeO<3\%$)

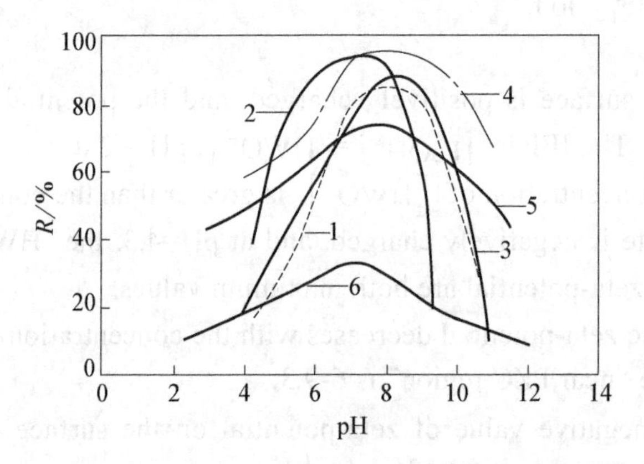

Figure 4-13 Relationship between wolframite flotation recovery and pH value (Collector dosage: mg /L)
1—sodium oleate 50; 2—styrene phosphonic acid 47; 3—8-hydroxyquinoline 200; 4—octyl hydroxamic acid 50; 5—1-nitroso-2-naphthol 50; 6—toluidine arsonic acid 50

When pH < IEP, the surface of wolframite is positively charged, but the collector is not dissociated, which is not conducive to the adsorption of anionic collector on the mineral surface by electrostatic force. When IEP < pH < 6, the wolframite surface has a high negative potential, and the potential determining ions are MOH^+ and HWO_4^-, which is not conducive to the adsorption of collector by electrostatic force and chemical force. $pH>pH_s$, the negative value of zeta potential on wolframite surface is very large, which is not conducive to the adsorption of anionic collectors. Therefore, only in $6<pH<pH_s$, that is, the "near PZC region", the negative value of zeta potential on the wolframite surface is small, and the potential determining ion is Me^{2+}, which is conducive to the adsorption of the anion collector by electrostatic force, and it can be chemically bonded to the collector through Me^{2+}. The result of this is that both the anionic collectors adsorbed by electrostatic force or the collectors chemisorbed by bonding with iron and manganese ions will have optimal flotation in the near PZC region of wolframite.

According to the same method as above, the logarithmic concentration diagram of dissolved components in the saturated scheelite solution can be drawn from equations (4-47) to (4-49) and equations (4-56) to (4-59) (see Figure 4-14). It can be predicted that the theoretical IEP of scheelite is 1.3, and the negative value of zeta potential rises to the maximum around pH=4.7. After that, it remained unchanged until pH=13.7, and the potential determining ions were Ca^{2+} and WO_4^{2-} ions. Figure 4-15 shows the relationship between the zeta potential and pH of scheelite reported in the literature, which is consistent with the predicted results.

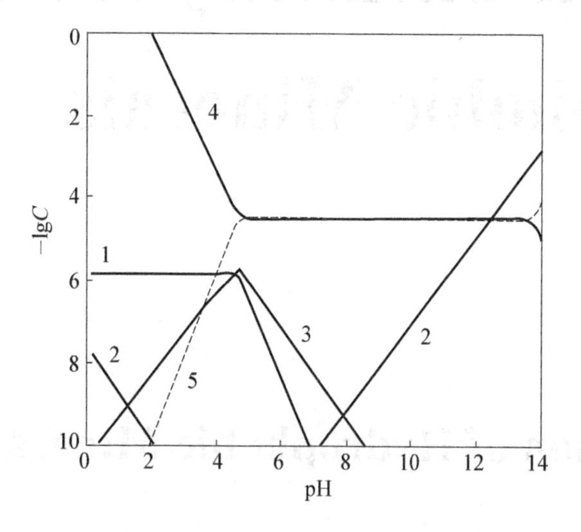

Figure 4-14 The logarithmic concentration diagram of solution components in a saturated solution of scheelite

1— $H_2WO_{4(s)}$; 2— $CaOH^+$; 3— HWO_4^- ; 4— Ca^{2+} ; 5— WO_4^{2-}

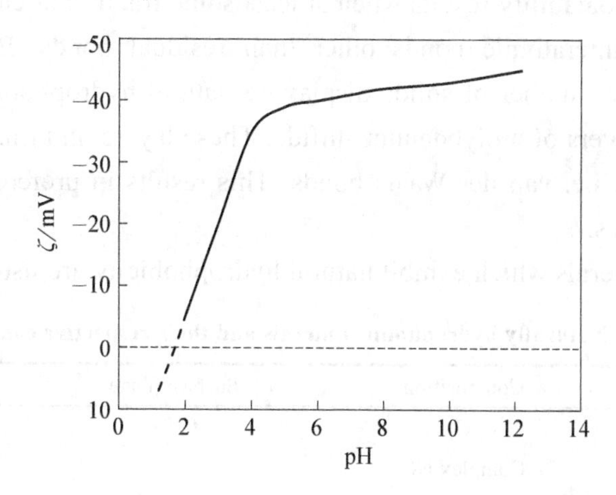

Figure 4-15 Relationship between zeta potential of scheelite and pH

In summary, the logarithmic concentration diagram of mineral dissolved components can analyze the structure of the electric double layer in the mineral surface and discuss the flotation mechanism.

Chapter 5
Flotation Chemistry of Naturally Hydrophobic Minerals

5.1 Flotation of Hydrophobic Minerals

5.1.1 Naturally hydrophobic minerals

Natural hydrophobicity of solids results principally from structural and bonding phenomena. Native floatability results when at least some fracture or cleavage surfaces form without rupture of interatomic bonds other than residual bonds. By way of example, molybdenite, one of a number of solids displaying natural hydrophobicity, is composed of electrically neutral layers of molybdenum sulfide. These layers, in turn, are held together by weak residual forces, i.e. van der Waals bonds. This results in preferential cleavage along the (0001) basal planes.

Some of the minerals which exhibit natural hydrophobicity are listed in Table 5-1.

Table 5-1 Naturally hydrophobic minerals and their respective contact angles

Mineral	Composition	Surface plane	Contact angle/(°)
Graphite	C	0001	86
Coal	Complex HC		20-60
Sulfur	S		85
Molybdenite	MoS_2	0001	75
Stibnite	Sb_2S_3	010	
Pyrophyllite	$Al_2(Si_4O_{10})(OH)_2$	001	
Talc	$Mg_3(Si_4O_{10})(OH)_2$	001	88
Iodyrite	AgI		20

Although the selected surfaces of these solids have a net hydrophobic character, two additional factors must be considered. First, whereas the overall behavior of the surface is classified as having a net hydrophobic character, the fact is that there may be a significant number of hydrophilic sites on the surface. As a result hydrophobic solids may still exhibit a surface charge and may have an adsorption potential for solutes arising from coulombic forces, chemisorption forces, and/or hydrogen bonding forces. As might be expected, the degree of hydrophobicity is greatest when the surface potential exhibits a minimum.

These phenomena are revealed in the electrokinetic responses of various molybdenite samples given in Figure 5-1. More negative potentials are realized as the edge/face ratio of the molybdenite is increased. The greater the contribution of the edges to the total area, the greater is the number of hydrophilic sites arising from oxidation and formation of thiomolybdate anion. Further, during size reduction crystal planes, other than those exhibiting a net hydrophobic character, may be exposed. As a consequence particles that might be considered to exhibit native floatability may, in fact, have a significant fraction of their surface consisting of other cleavage planes which do not exhibit a hydrophobic character.

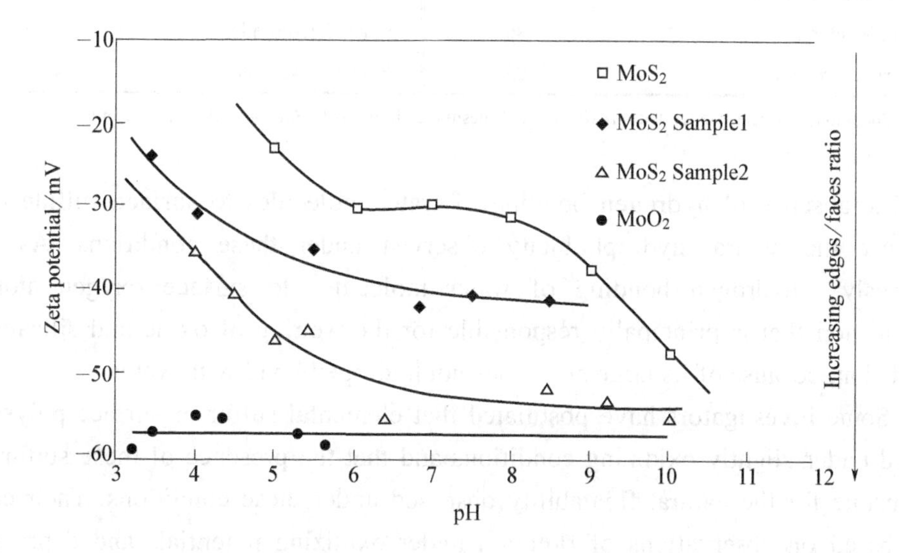

Figure 5-1 Zeta potential of various samples of molybdenite and molybdic oxide as a function of pH

5.1.2 Sulfides in oxygen-deficient systems

Flotation of various sulfides in the virtual absence of oxygen, i.e. in water containing less than 5×10^{-9} oxygen, has shown the natural floatability of these minerals under these conditions. See Table 5-2. As can be noted, essentially complete flotation of galena, chalcopyrite, chalcocite and pyrite occurs under these conditions. In the case of sphalerite, the surface exhibits about the same degree of hydrophobicity as hydrophilicity.

Following copper activation, complete flotation of sphalerite occurs.

Table 5-2 Flotation recovery of sulfides from various sources①

Mineral	Recovery/%	Mineral	Recovery/%
Galena		**Pyrite**	
Coeur D'Alene, Idaho	100	Ambo Segues, Spain	92
Bixby, Missouri	100	Custer, South Dakota	85
Pitcher, Oklahoma	100	Zacatecas, Mexico	83
Galena, South Dakota	100	Naira, Mexico	82
Chalcopyrite		**Sphalerite**	
Temagami, Ontario	100	Keystone, South Dakota	56
Sudbury, Ontario	100	Joplin, Missouri	47
Beaver Lake District, Utah	97	Greede, Colorado	46
Messina, Transvaal	93	Pitcher, Oklahoma	41
Chalcocite		**Sphalerite (Cu^{2+} activated)**	
Kennecott, Alaska	100	Keystone, South Dakota	100
Evergreen, Colorado	88	Joplin, Missouri	100
Butte, Montana	86	Graced, Colorado	100
Superior, Arizona	83	Pitcher, Oklahoma	100

① Conditions: particle size, 100 mesh×200 mesh (1 mesh=0.254mm); pH = 6.8, no collector, no frother, oxygen free.

The absence of hydrogen bonding of water molecules to surface sulfide atoms may influence the natural hydrophobicity observed under these conditions. As mentioned previously, hydrogen bonding of water molecules to surface oxygen atoms is the phenomenon that is principally responsible for the wetting of oxide and silicate minerals. Sulfide ion, because of its large size, does not hydrogen bond with water.

Some investigators have postulated that elemental sulfur or surface polysulfides are formed under slightly oxidizing conditions and that the presence of these sulfur species is responsible for the natural floatability observed under these conditions. Their conclusions were based on observations of flotation under oxidizing potentials and depression under reducing conditions.

The general area of natural floatability of sulfides is important from the standpoint of mechanisms involved in sulfhydryl collector adsorption as well as practical flotation separations. Clear understanding of these phenomena has not been achieved as yet, however, further work is warranted.

5.1.3 Flotation and depression of naturally hydrophobic minerals

For natural hydrophobic minerals, such as coal, graphite, molybdenite and other non-polar minerals, non-polar hydrocarbon oil is used as collector during flotation, assisted by frothers and promoters. These non-polar hydrocarbon oils generally come from petrochemical industry, such as kerosene, light diesel oil, etc., or other chemical by-products. The main components of non-polar hydrocarbon oils are aliphatic hydrocarbons, cycloalkanes and aromatic hydrocarbons. Their molecules are composed of carbon and hydrogen atoms, which are combined by non-polar carbon carbon bonds and weak polar carbon hydrogen bonds. Molecular bonds are held together by dispersive forces. Oil is insoluble in water, and non-polar hydrocarbon oil can only exist as oil droplets in water. During the flotation of natural hydrocarbon minerals, due to their non-polar surface, hydrocarbon oil can spread on the surface of these minerals, forming a hydrocarbon oil film on the surface, which improves the hydrophilicity of these solid surfaces. For other polar minerals, the attraction of the mineral surface to oil molecules is less than that to water. The mineral surface is covered with water molecules to form a hydration film, and oil molecules cannot be adsorbed on the mineral surface. Non-polar oils can improve the adhesion of hydrophobic minerals and air bubbles, and increase the degree of adhesion.

There are some hydraulic polar parts on the surface of hydraulic minerals. When the proportion of polar parts increases, the flotation effect can be improved by adding a flotation promoter. Promoter is a flotation reagent that can improve the effect of collector and frother. Its composition is mostly surface active organic compounds, and it is an important auxiliary agent for coal flotation. Most promoters are not collectors and frothers themselves, but adding a small amount of them can improve the performance of the agent, improve the flotation effect, increase the yield of concentrate, and reduce the consumption of flotation reagent.

Emulsifier, considered by some as an accelerator, can improve the effect of collector and frother. It is usually a surfactant with heteropolar molecular structure. The emulsifier makes the hydrocarbon oil easily dispersed into small oil droplets in pulp by emulsifying the hydrocarbon oil. It will also disperse the frother and play an adjustment role. Emulsifying the agent with appropriate methods can greatly improve the dispersibility of the agent, reduce the dosage of the agent, reduce the adhesion of fine mud on the surface of the flotation concentrate, and improve the quality of the concentrate. There are many methods of pharmaceutical emulsification, such as ultrasonic emulsification and emulsifier emulsification.

Depression of solids which exhibit natural floatability can be achieved by chemical alteration of the surface. Usually, extensive oxidation is required. Alternatively, depression can be achieved by the adsorption of organic colloids, specifically, derivatives of starch. See Figure 5-2. Apparently these polymers bond hydrophobically to the mineral surface and extend their polar hydroxyl groups to the aqueous phase so that water molecules are oriented in the polar force field. The naturally hydrophobic mineral then becomes hydrophilic. The non-specific nature of this bonding is revealed by the fact that the same dextrin adsorption isotherm fits three naturally hydrophobic minerals of quite different chemical composition as shown in Figure 5-2. Furthermore, the heat of adsorption is the same in each case, namely, about −0.5 kcal/mol of monomeric unit.

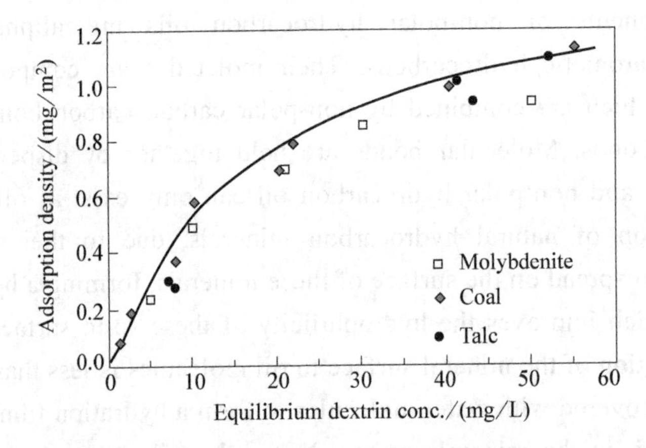

Figure 5-2 Adsorption isotherm of dextrin on various naturally hydrophobic minerals

5.2 Flotation Practice of Coal

5.2.1 Floatability of coal

Although coal exhibits natural floatability, its response to flotation varies with carbon content and ash content. As shown in Figure 5-3 and in Table 5-3, the maximum contact angle occurs for low volatile bituminous and semi-anthracite coals. Further, the contact angle, and hence flotation, of the highest rank coal, anthracite, is somewhat less.

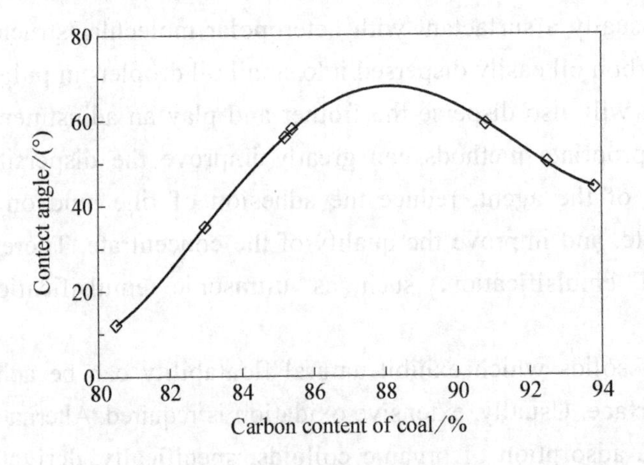

Figure 5-3 Variation of contact angle with carbon content of coal

With regard to ash content, the greater the ash content of a coal, the less hydrophobic is the coal. The mineralogical composition of ash constituents of coal is given in Table 5-4, and it can be noted that ash consists principally of silicates. Since silicates are readily wetted by water, the reduction of natural hydrophobicity with increasing ash content would be expected.

Minerals in coal are the general term of inorganic substances in coal, including minerals independently existing in coal, such as clay minerals, pyrite, calcite, quartz, etc. It also includes inorganic elements combined with the organic matter of coal, which exist in the form of carbonyl salts, such as calcium, sodium, etc. Clay mineral is the most important mineral in coal, and its content is significantly higher than that of other minerals.

Sulfur content of coal is of considerable importance from an environmental standpoint. Two forms of sulfur are present, namely, organic sulfur and inorganic sulfur. The organic sulfur is incorporated structurally in the macromolecular coal polymer as thiophenes, sulfur ring structures, which cannot be separated by flotation techniques. The inorganic sulfur is present largely as ferrous disulfide in the form of pyrite and marcasite. Occasionally, gypsum contributes to the inorganic sulfur component. Generally, this form of sulfur can be separated from coal although recent data suggest that significant quantities of colloidal pyrite (−1μm in size) are present which may preclude separation.

Table 5-3 Approximate values of some coal properties in different rank range

			Bituminous					Anthracite
			High vol.					
	Lignite	Subbit.	C	B	A	Medium vol.	Low vol.	
C (min, matter free)/%	65-72	72-76	76-78	78-80	80-87	89	90	93
H/%	4.5	5	5.5	5.5	5.5	4.5	3.5	2.5
O/%	30	18	13	10	10-4	3-4	3	2
O as COOH/%	13-10	5-2	0	0	0	0	0	0
O as OH/%	15-10	12-10	9	?[1]	7-3	1-2	0-1	0
Aromatic C atoms of total C/%	50	65	?[1]	?[1]	75	80-85	85-90	90-95
Av. no. benz. Rings per layer	1-2	?[1]	←	2-3	→	→	5?[1]	>25?[1]
Volatile matter/ %	40-50	35-50	35-45	?[1]	31-40	31-20	20-10	<10
Reflectance<vitrinite/%	0.2-0.3	0.3-0.4	0.5	0.6	0.6-1.0	1.4	2.8	4
Density	→	→	→	→	→	increases	→	→
Total surface area	←	←	←	←	←	minimum	→	→
Plasticity and coke formation			|←	←	←	only	→|	
Calorific value, moist, min, matter free/(Btu/lb).	7,000	10,000	12000	13500	14,500	15,000	15,800	15,200

Note: 1Btu=1055.06J, 1lb=0.4536kg.

[1] Not obtained or inaccurate.

Table 5-4 Average mineralogical composition of as constituents of U.S. coal seams

Mineral	Standard/%		
	Mean	Deviation	Range
Kaolinite	34.8	23.6	0-85
Illite	7.8	1.5	0-36
Montmorillonite	0.7	1.3	0-10

(continued)

Mineral	Standard/%		
	Mean	Deviation	Range
Mixed layer illite –montmorillonite	3.2	2.3	0-20
Chlorite	1.5	1.5	0-10
Quartz	10.1	10.1	0-40
Gypsum	11.9	13.0	0-60
Rutile	2.3	1.3	0-10

Note: Others include pyrite, siderite, dolomite, calcite, aragonite, ankerite, muscovite, plagioclase, hematite, jarosite, thenardite.

5.2.2 Coal flotation

(1) Decarbonization flotation. Coal flotation is mainly to reduce the ash content of coal. Although coal exhibits natural hydrophobicity, its surface exhibits hydrophilic sites, and usual flotation practice involves the use of frother (NIBC, pine oil or acetylic acid) on the order of 0.2lb/ton and a promotor (fuel oil) on the order of 1-3lb/ton depending on the coal. The latter addition is made to mask hydrophilic sites arising from the heterogeneous nature of the solid. With this flotation strategy, a major portion (~85%) of the feed material is to be recovered in the concentrate. As a result considerable mechanical entrainment of gangue can occur in the froth phase, and the quality of the separation may be impaired.

At present, coal flotation is mainly used to recover coal from low rank coal and slime and gangue produced by gravity separation. Among them, the use of flotation to recover slime from heavy medium coal preparation is widely used. The flotation of coal slime is relatively difficult for the following reasons:

① Coal particles must overcome the hydration layer on the surface of the coal particles and the circulation effect on the surface of the bubbles to stick to the bubbles. Due to the small mass of the fine particles, it is difficult for the hydrophobic ore particles to adhere to the surface of the bubbles to form mineralized bubbles.

Fine-grained gangue is highly viscous by the water medium, and once it adheres to the bubbles, it is difficult to fall off, and the "secondary enrichment" effect is weakened, and it is very easy to be mixed between the mineralized bubbles and rise into the foam layer to form entrainment. As a result, the concentrate that should enter the foam layer does not all enter the foam layer, and the argillaceous high-ash minerals that should not enter the foam layer enter the foam layer, resulting in a decrease in the quality of clean coal, that is, flotation selectivity is poor.

② High surface energy results in serious non-selective agglomeration between gangue ore particles and useful ore particles, enhanced surface hydrophilicity, increased adsorption capacity, and a corresponding decrease in selectivity. This is the main reason for the poor flotation effect of fine-grained minerals.

③ Due to the existence of a large number of non selective attachments, argillaceous minerals are attached to the bubble surface, which increases the viscosity and stability of foam. Not only does the "secondary enrichment" function weaken, the ash content of fine-grained clean coal increases, and the flotation selectivity becomes poor, but also it causes certain difficulties for the next step of product dehydration.

④ Non polar collectors such as kerosene and light diesel oil are usually used for coal flotation. Due to the strong hydrophobicity on the surface of the oil collector, the collector has poor dispersion performance in the slurry mixing cylinder and cannot form a fine dispersed phase, which reduces the probability of contact between the oil collector and the mineral particles and reduces the degree of mineralization of the mineral particles.

Therefore, in addition to adding collectors and foaming agents before slime flotation, different regulators, such as accelerators, pH regulators, and inhibitors, should be added according to the situation. As an auxiliary agent to control the interaction between minerals and collectors, the rational use of modifiers is the key to obtaining high separation.

(2) Desulfurization flotation. There are also reports of desulfurization from coal by flotation. Generally, pyrite inhibitors are added when floating coal is used to achieve sulfur removal.

An alternative flotation strategy has been proposed. This two-stage process for desulfurization involves flotation of pyrite from the depressed coal. The first stage is conventional flotation, while the second stage involves conditioning the first stage concentrate with a dextrin-type depressant and a sulfhydryl collector. The pyrite is floated from coal in the second stage.

5.3 Flotation Practice of Molybdenite

5.3.1 Floatability of molybdenite

The floatability of molybdenite minerals mainly depends on the type of broken bonds on the surface of the mineral particles. Low surface energy surface formed by molecular bond breaking(molybdenite [001] plane, 2.4×10^{-2} J/m^2). It is mainly acted by van der Waals force, with weak surface polarity and strong hydrophobicity. On the contrary, the high energy surface formed by covalent bond or ionic bond breaking (on molybdenite edge, 0.7 J/m^2). It is mainly affected by ionic bonding force, with strong surface polarity and weak hydrophobicity. Molybdenite is a typical heteropolar mineral. The floatability of molybdenite is closely related to the ratio of edge and face of particles. Non-polar oil collectors are only suitable for adsorbing non-polar surfaces of minerals that are broken by molecular bonds.

The large area ratio between edges and surfaces of fine molybdenite reduces the adsorption of non-polar oil on its particles, which is the reason for the low flotation recovery of fine molybdenite when non-polar oil is used as collector. The relationship between the floatability of molybdenite and its particle size is shown in Figure 5-4. Studies have shown that molybdenite with particle size of 10-100μm has high surface to edge ratio, strong hydrophobicity and good floatability. The surface to edge ratio of molybdenite with particle size of ~10μm decreases sharply, resulting in poor flotation behavior and difficult recovery.

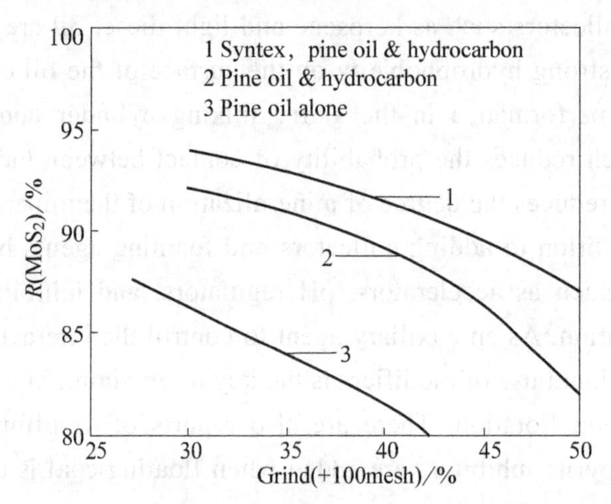

Figure 5-4 Relationship between recovery of molybdenite and grind for various reagent conditions

5.3.2 Molybdenite flotation

In the flotation practice, the floatability of molybdenite is also related to the embedded particle size of molybdenite, the oxidation rate of molybdenite surface, the types of co-existing sulfide minerals and the types of associated gangue minerals.

Production of molybdenite by froth flotation results from the processing of primary molybdenum ores and copper porphyry ores in which molybdenite is recovered as a by-product. Approximately one third of the total molybdenum production in the western world comes from by-product molybdenite.

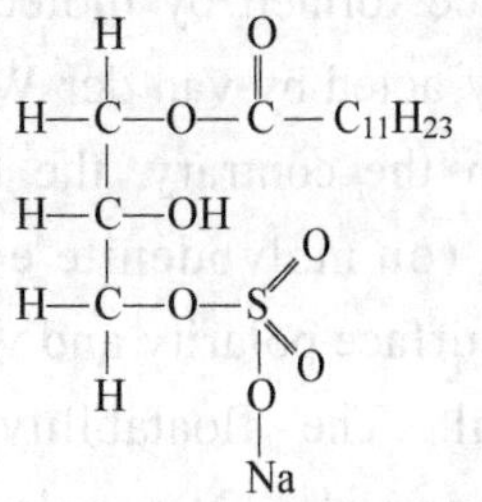

The initial stage of flotation of primary molybdenum ores is accomplished at a relatively coarse grind. At Climax Molybdenum, for example, the first separation is made at 35 percent +100 mesh. Vapor oil is added as a promoter, and pine oil is added as frother. Emulsification of the oil is effected by the addition of a sulfated coconut oil (Syntex) to the system. The structure of this reagent is shown in Figure 5-5. The beneficial effect of each of these is illustrated in Figure5-4.

H—C—O—C(=O)—$C_{11}H_{23}$
H—C—OH
H—C—O—S(=O)(=O)—O—Na
H

Figure 5-5 The structure of Syntex

The processing of copper-molybdenum concentrates obtained from copper porphyry ores by flotation techniques involves rendering either molybdenite or the copper sulfide minerals hydrophilic. Typical bulk copper-moly concentrates contain 25 percent copper and, frequently, less than 1 percent molybdenum. Generally, the copper sulfides are made hydrophilic by conditioning the bulk copper-moly concentrate with specific copper sulfide depressants. Copper sulfide depression rather than molybdenite depression is selected because, in so doing, the minor constituent MoS_2, is floated with a resultant lower probability of contamination due to mechanical transport of unwanted material to the froth phase. In a few instances, molybdenite in depressed by dextrin, and the copper sulfides are floated, but this is the exception rather than the rule.

Molybdenite has excellent natural floatability, and chalcopyrite also has good floatability. Generally, the combination of hydrocarbon oil and xanthate is used for mixed flotation of copper and molybdenum. In this case, a large number of xanthate collectors will be adsorbed on the surface of copper sulfide to form various hydrophobic substances, which makes it more difficult to inhibit copper sulfide, resulting in poor separation effect of copper and molybdenum.

The selection of appropriate copper mineral inhibitors plays a decisive role in improving the separation index of copper and molybdenum. A variety of reagents can be used to effect depression of copper sulfide minerals from bulk copper-moly concentrates:

① Alkali sulfides and polysulfides

② Noke's type reagents (group V-A polysulfides)

③ Oxidants (both thermal and chemical treatments)

④ Cyanide, ferrocyanide, ferricyanide

⑤ Organic inhibitors(mainly include mercaptoacetic acid compounds, thiourea, carboxylic acid, sulfonic acid, thiocarbonate, etc.)

Without exception, combinations of the above mentioned depressants are used industrially. Furthermore, in many cases the copper-moly concentrate is heat treated either by steaming or roasting prior to the copper-moly separation. Oxidation and destruction of collector may be accomplished by both techniques.

Chapter 6
Flotation Chemistry of Sulfide

6.1 General Considerations in Xanthate Adsorption

6.1.1 Oxygen-deficient system

Adsorption studies have been conducted in aqueous solutions containing less than 5 parts per billion oxygen. Adsorption of various xanthates on galena in this system is presented as a function of pH in Figure 6-1. Two phenomena can be noted, namely, that adsorption is independent of pH and, secondly, that adsorption density is some what less for longer-chain xanthates. The adsorption density of 2×10^{-6} mole EX^{-} per gram galena represents monolayer coverage assuming one xanthate ion adsorbs on one surface lead ion site. The fact that adsorption occurs at high values of pH indicates that xanthate ion rather than xanthic acid is the active species in this system.

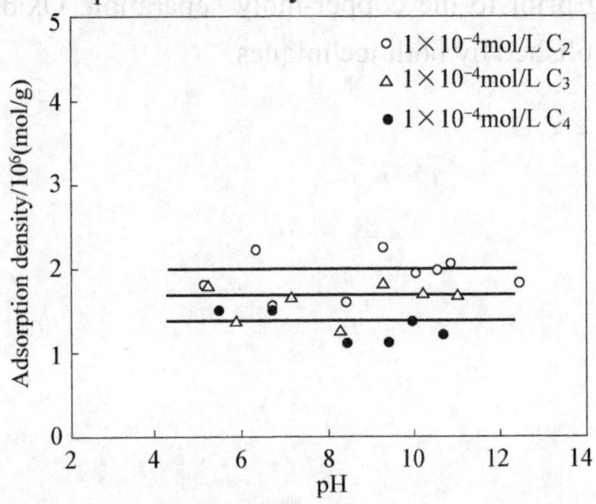

Figure 6-1 Adsorption density of various xanthates on galena as a function of pH in the virtual absence of oxygen (Surface area of galena: 0.243 m^2/g)

Chemisorption of xanthate under these conditions can be noted from the data in Figure 6-2 in which the zeta potential of galena in the virtual absence of oxygen is presented as a function of pH. It can be noted that galena has a pzc at pH 2.6 under these conditions and that the zeta potential is negative above this pH. In the presence of 1×10^{-3} mol/L ethyl xanthate, the zeta potential becomes more negative over this pH range. Similarly to the adsorption data presented in Figure 6-1, the zeta potential is constant and independent of pH. An increase in the zeta potential is indicative of the formation of a new phase on the mineral surface. Further, adsorption of a negatively-charged, short-chain collector ion on a negatively charged-surface could occur only by chemical interaction between the collector and surface metal ion.

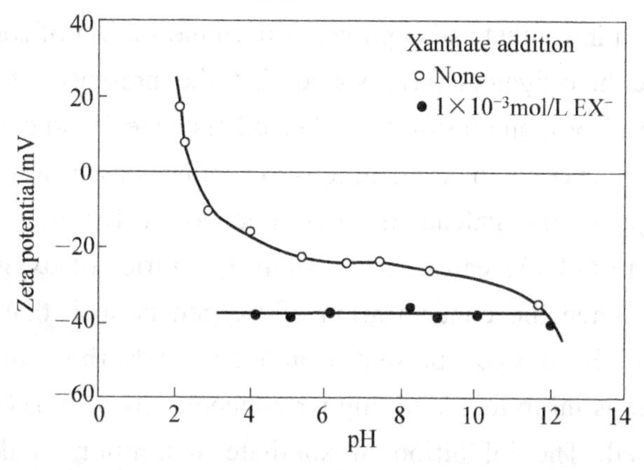

Figure 6-2 Zeta potential of galena as a function of pH in the absence and presence of ethyl xanthate in the virtual absence of oxygen

The significant difference in surface properties of galena in the absence and presence of oxygen can be seen by comparing the electro-kinetic data given in Figures 6-2 and 6-3. The data in Figure 6-3 were established in the presence of air, and a point-of-zero-charge can be noted at pH 7.9. This can be contrasted with a pzc of pH 2.6 in the virtual absence of oxygen. Chemisorption of xanthate can also be noted from these data.

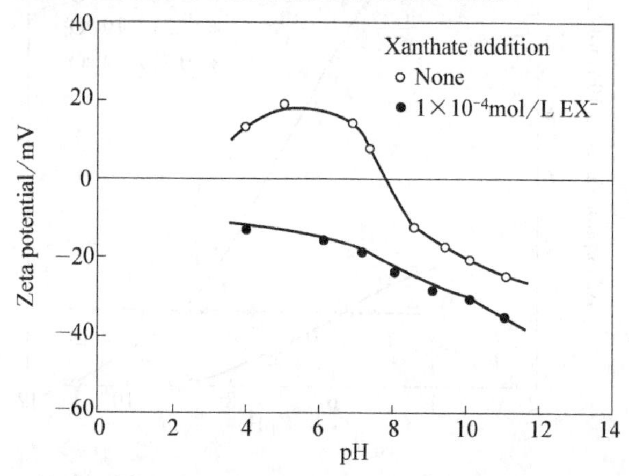

Figure 6-3 Zeta potential of galena as a function of pH in the absence and presence of ethyl xanthate in the presence of oxygen

Adsorption of various xanthates on sphalerite from 1×10^{-4} mol/L xanthate solutions occurs only below monolayer coverage. Adsorption is not possible above pH 6.6 with heptyl xanthate, above pH 7.4 with octyl xanthate and pH 10.3 with dodecyl xanthate. Electrophoretic determinations revealed that ethyl xanthate is also chemisorbed on sphalerite under these conditions. In the case of pyrite, adsorption of short-chain xanthates does not occur above about pH 8 in the virtual absence of oxygen, and adsorption is below monolayer coverage. Similarly to sphalerite, ethyl xanthate chemisorbs on pyrite under these conditions.

6.1.2 Presence of oxygen

The role of oxygen in xanthate adsorption has been the subject of considerable controversy over the years. Some investigators have stated that the presence of oxygen is absolutely necessary for xanthate adsorption on sulfides, while others have found the presence of quantities of oxygen at monolayer coverage to be deleterious to xanthate adsorption.

Xanthate adsorption on galena is presented as a function of pH and oxygen concentration in Figure 6-4. As can be seen, small quantities of oxygen have no effect on xanthate adsorption. When the concentration of oxygen is such that monolayer coverage would result if all of the oxygen present would have adsorbed, in this case 0.31×10^{-6}, adsorption of xanthate is inhibited. With higher concentrations of oxygen, though, xanthate adsorption is enhanced. The inhibition of xanthate adsorption in the presence of small quantities of oxygen is kinetically controlled. The data contained in Figure 6-4 were established with a 5-minute conditioning period. With a 10-minute conditioning period, xanthate adsorption is enhanced in the presence of oxygen. In fact, multiplayer coverage of xanthate is obtained under these conditions.

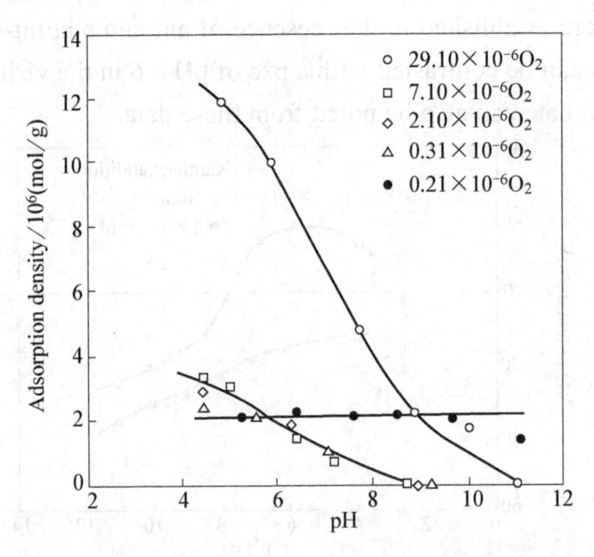

Figure 6-4 Adsorption density of ethyl xanthate on galena as a function of pH and oxygen concentration

Xanthate concentration: 1×10^{-4} mol/L, surface area of galena: 0.243m^2/g

In the cases of pyrite and sphalerite, any amount of oxygen present in the system enhances xanthate adsorption even with relatively short conditioning periods. In the presence of oxygen, oxidation of the surface sulfide ion to sulfur-oxy species occurs. Surface analyses by ESCA (electron spectroscopy for chemical analysis) and Auger spectroscopy have provided direct experimental evidence of this phenomenon. Under these conditions water will be hydrogen bonded to the surface, and the mineral will lose its natural floatability. In the case of molybdenite, however, this mineral is extremely refractory to oxidation. As a result, it possesses natural floatability even in the presence of air.

Collectors must be added to float most metal sulfides in the presence of air, then, and short-chain sulfhydryl collectors are commonly employed. The fact that collectors with as few as two carbon atoms in the hydrocarbon chain may be used is due to a number of phenomena. First, although surface sulfide ion is readily oxidized by dissolved oxygen, not all of the sulfide ion sites will have been oxidized during the time that solids' preparation and flotation are conducted. As a result, these minerals will be much less hydrophilic than oxides and, hence, much easier to float. Secondly, insoluble salts form between heavy metal ions and short-chain sulfhydryl collectors at relatively low concentration. Finally, certain sulfhydryl collectors undergo oxidation to dimers such as dixanthogen and dithiophosphatogen at the surface of certain sulfide minerals.

Mechanisms by which sulfhydryl-collector species adsorb involve electrochemical phenomena in some mineral systems and chemisorption in others.

6.1.3 Electrochemical phenomena in sulfide systems

As discussed in previous sections, it has been observed that under anaerobic conditions sulfide minerals are hydrophobic. Extensive surface hydration of sulfide minerals, leading to a hydrophilic character, occurs on exposure to small oxygen potentials. The sensitivity of sulfide minerals to oxygen is further revealed from rest potential measurements at a freshly cleaved galena surface. Semiconducting sulfide minerals, such as galena, develop a rest potential when placed in an aqueous solution. The rest potential for a galena electrode before cleavage was found to be −0.21 volt (S.C.E.) as shown in Figure 6-5. Upon cleavage the potential drops almost instantaneously to −0.615 volt (S.C.E.) and then gradually is restored to its original value after several hundred seconds. During this time the surface reacts with oxygen present in the system. These experimental results clearly indicate from an electrochemical viewpoint the significant difference to be expected in the surface state of sulfide minerals on exposure to oxygen. In most cases, even with a blanket of tank nitrogen, there is sufficient oxygen present in the system to alter the rest potential of sulfide minerals from their intrinsic values. Nevertheless, these rest potentials seem to be reproducible. Rest potentials of various sulfide minerals have been established under flotation conditions and are reported in Table 6-1 for a solution containing 6.25×10^{-4} mol/L ethyl xanthate at pH 7. Comparison of these values should be made with the reversible potential for xanthate oxidation in this system which is 0.13 volt (IUPAC). This value is obtained as follows:

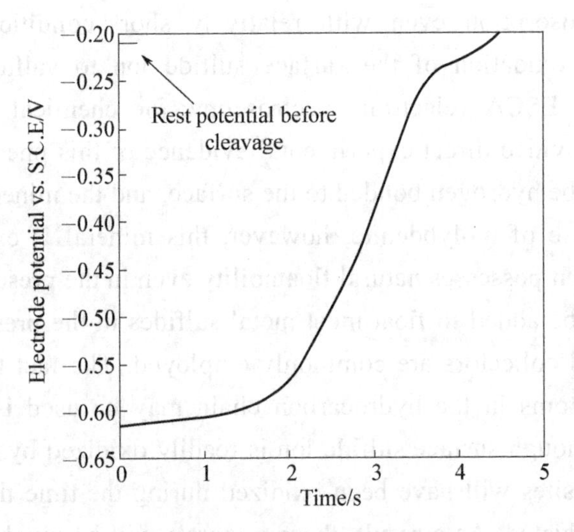

Figure 6-5 Electrode potential of galena as function of time after cleavage under nitrogen-purged 0.1mol/L borate solution

$$X_{2(l)} + 2e^- \rightleftharpoons 2X^- \quad E^0 = -0.06\text{V} \tag{6-1}$$

$$E_{rev} = E^0 - \frac{RT}{nF}\ln\frac{\left(X^-\right)^2}{\left[X_{2(l)}\right]} \tag{6-2}$$

$$E_{rev} = -0.06 - \frac{1.98\times 298}{2\times 23060}\ln\frac{\left(6.25\times 10^{-4}\right)^2}{(1)} \tag{6-3}$$

$$E_{rev} = -0.06 + 0.19 = +0.13(\text{V}) \tag{6-4}$$

Since dixanthogen, $X_{2(l)}$, is a pure liquid, its activity is unity.

When the rest potential is larger than the reversible or Nernst potential, oxidation of xanthate to dixanthogen occurs. With reference to Table 6-1, dixanthogen is the reaction product found on the various mineral surfaces with rest potentials greater than 0.13 volt. Covellite appears to be an exception, but these anomalous results are apparently due to dissolution of the covellite followed by oxidation of xanthate with Cu^{2+} and adsorption of dixanthogen from solution.

When the rest potential is cathodic or less than the reversible xanthate/dixanthogen potential, oxidation of xanthate cannot occur, and metal xanthates are observed on the sulfide surface.

Table 6-1 Products of the interaction of sulfide minerals with ethyl xanthate (6.25×10^{-4} mol/L at pH 7)

Mineral	Rest potential/V	Product
Pyrite	0.22	Dixanthogen
Arsenopyrite	0.22	Dixanthogen

(continued)

Mineral	Rest potential/V	Product
Pyrrhotite	0.21	Dixanthogen
Chalcopyrite	0.14	Dixanthogen
Covellite	0.05	Dixanthogen
Bornite	0.06	Metal xanthate
Galena	0.06	Metal xanthate
Chalcocite	0.06	Metal xanthate

Note: Reversible potential for oxidation to dixanthogen is 0.13V.

6.1.4 Effect of semiconductor property of sulphide mineral on xanthate adsorption

Most sulphide minerals are semiconductors as shown in Table 6-2. The difference of interactions of sulphide minerals with a collector at certain conditions is originated from their different semiconductor properties. For example, it was reported that p-type galena adsorbed xanthate more easily. The increase of the hole on the galena surface after light irradiation promoted the adsorption of xanthate. The pyrite in ore is p-type and the pyrite in coal is n-type giving rise to their different flotation behavior. n-type pyrite-band was helpful for the reduction of oxygen. The change of the electronic structure (e.g. Femi energy level) of galena and pyrite surface with pH and the concentration of reagent noted that the interaction between mineral and reagent depended on their energy match. Most importantly, the flotation behavior of sulphide minerals was reported to be able to be controlled by changing the surface electronic structure or energy level through light irradiation and other methods.

Table 6-2 Semiconductor properties of different sulphide minerals

Minerals	Width of forbidden ban (eV) or conductivity	Minerals	Width of forbidden ban (eV) or conductivity
PbS	0.41	Sb_2S_3	1.72
ZnS	3.6	As_2S_3	2.44
(Zn,Fe)S①	0.49	HgS	2
$CuFeS_2$	0.5	CoS_2	<0.1
FeS_2	0.9	CuS	Meta conductor
NiS_2	0.27	CoS	Meta conductor
Cu_2S	2.1	FeS	Meta conductor

① The mass fraction of Fe—ω(Fe)=12.4%.

The authors used CNDO/2 (complete neglect of differential overlap/version 2) method of quantum chemistry and band theory to study the flotation mechanism of sulphide minerals in the presence and absence of collectors.

Figure 6-6 shows the energy level diagram of molecular orbital of galena, pyrite, common and activated oxygen and ethyl xanthate ion. The highest occupied molecular orbital (HOMO) of galena surface mainly consists of 3p orbital of sulphur atoms, and the lowest unoccupied molecular orbital (LUMO) mainly consists of the 6p orbital of the lead atom. There is a little overlap between the 6p orbital of the lead atoms and the 3p orbital of the sulphur atoms. It indicates that the Pb-S bond at the galena surface contains a large ionic bond character and a small covalent bond character. In the case of pyrite, the atomic orbitals of iron and sulphur have equal contribution to the HOMO and LUMO. There is a great overlap between the atomic orbitals of iron and sulphur, showing that Fe-S bond exhibits covalent bond character mostly and the Fe-Fe bond exhibits the metal bond character partly. As above, surface electron structure is much different between the mineral of the type galena and that of the type pyrite. Oxygen possesses the following molecular orbital $(\sigma_{1s})^2(\sigma_{1s})^2(\sigma_{2s})^2(\sigma_{2s}^*)^2(\sigma_{2px})^2(\pi_{2py})^2(\pi_{2pz})^2(\pi_{2pz}^*)^1(\pi_{2py}^*)^1(\sigma_{2px}^*)^0$.

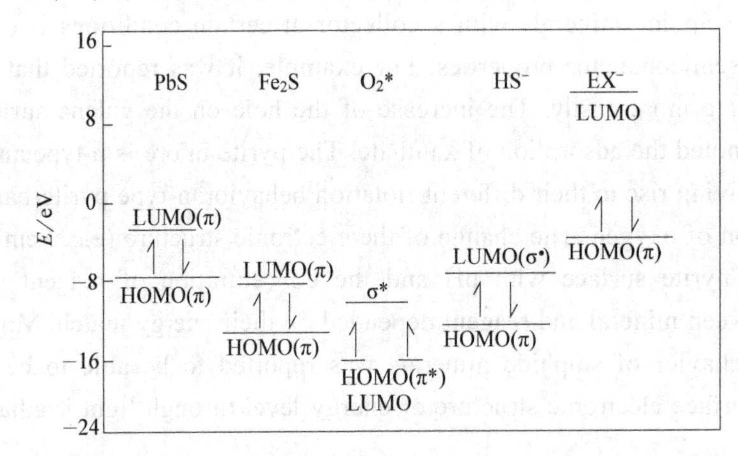

Figure 6-6 Symmetry and energy of frontier molecular orbital (HOMO and LUMO)

In its two anti-π orbital(π*), there are two single electrons. The orbital can either accept or donate one electron, which can be either HOMO or LUMO. From Figure 6-6, the energy values and symmetry of the frontier molecular orbital of galena, pyrite, HS^-, EX^- and oxygen are shown in Table 6-3.

Table 6-3 The energy values and symmetry of the frontier molecular orbital of galena, pyrite, HS^-, EX^- and oxygen

Reactants	Galena	Pyrite	Common oxygen	Activated oxygen	HS^- ion	Ethyl xanthate ion
Energy HOMO	−7.76	−13.27	−15.82	−12.60	−11.70	−3.70
Symmetry	π	π	π	π	π	π
Energy LUMO	−3.133	−9.41	−15.82	−12.60	−8.7	−12.1
Symmetry	π	π	π	π	σ	σ

6.2 Flotation Chemistry of Galena

6.2.1 Galena flotation

The study that galena has received over the years has been extensive. This is due in part to its availability in natural form and also because it does not have the complicating effect of containing a metal ion that can change oxidation states readily such as copper or iron. The flotation response of galena with 1×10^{-5} mol/L ethyl xanthate in the presence of air is presented in Figure 6-7. As can be noted, complete flotation is effected from pH 2 to pH 10.

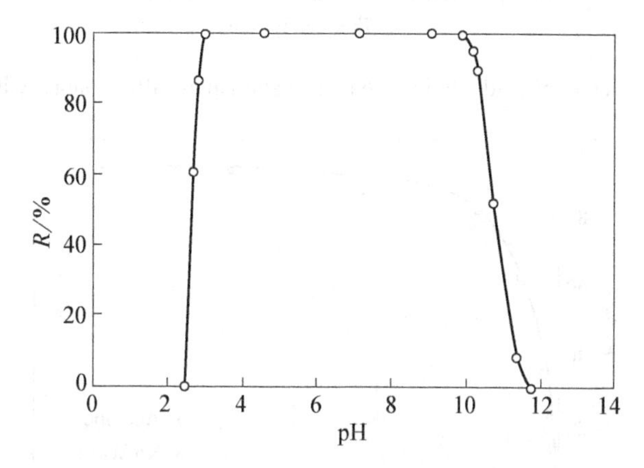

Figure 6-7 Flotation recovery of galena as a function of pH with 1×10^{-5} mol/L ethyl xanthate in the presence of air

Xanthate is present in two forms on a galena surface under those conditions in which flotation occurs. The one form is xanthate chemisorbed at monolayer coverage, while the other is bulk precipitated lead xanthate adsorbed at multiplayer coverage. Dixanthogen does not form under these conditions and is not observed. See Table 6-1. Confirmation of the presence of bulk lead xanthate on the galena surface is provided by the infrared spectra in Figure 6-8. Presented are the spectra of lead ethyl xanthate and galena after contact with ethyl xanthate. The absorption bands of lead ethyl xanthate occur at 1014 cm^{-1}, 1110 cm^{-1} and 1212 cm^{-1}. After contact with ethyl xanthate, absorption bands at 1020 cm^{-1}, 1112 cm^{-1} and 1210 cm^{-1} are noted.

The multilayers of lead ethyl xanthate are held together by van der Waal's bonding of the hydrocarbon chains of the xanthate, and these layers can be dissolved with organic reagents such as acetone. However, the xanthate chemisorbed at monolayer coverage cannot be leached from the surface. These facts can be noted from the data presented in Figure 6-9. Also shown in Figure 6-9 is the floatability of galena as a function of multilayer coverage of lead ethyl xanthate. For flotation to be effected in the presence of multilayer coverage, the precipitate of lead ethyl xanthate must be hydrophobic which has been shown to be the case.

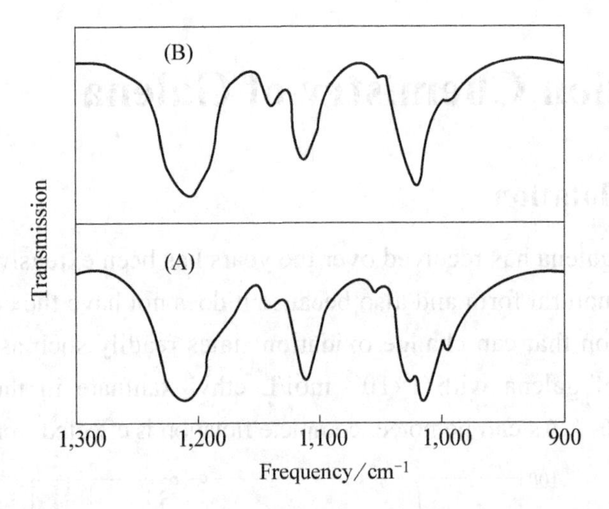

Figure 6-8 Infrared spectra of lead ethyl xanthate (A) and galena after contact with ethyl xanthate (B)

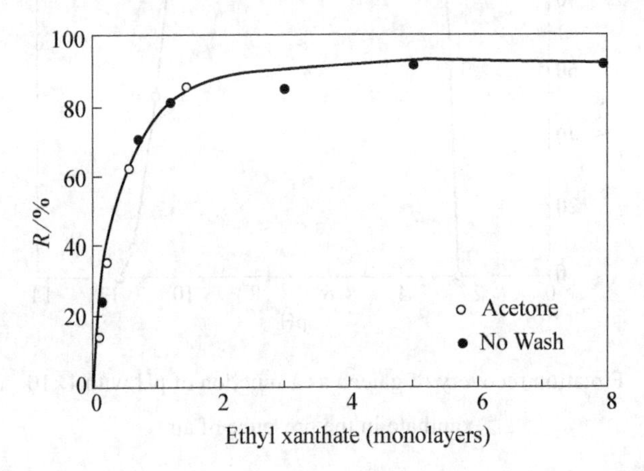

Figure 6-9 Variation in flotation of galena with concentration of xanthate at the surface

In the chemisorption of xanthate at monolayer coverage, one xanthate ion adsorbs on each surface lead ion to form an unleachable phase of lead xanthate. Some investigators have suggested that surface sulfide ion is oxidized electrochemically to elemental sulfur by oxygen in this adsorption process, but no elemental sulfur has ever been detected on the surface.

Electrochemical measurements with a galena electrode suggest that monolayer adsorption can involve charge transfer with the discharged xanthate ion being fixed at the galena surface.

$$EX^- = EX_{ads} + e^- \quad (6\text{-}5)$$

The voltammograms shown in Figure 6-10 present the current density/potential relationship for a galena electrode in the presence and absence of ethyl xanthate. In both cases the electrode behaves reversibly. The anodic curve in the presence of ethyl xanthate exhibits a peak at about 0 volts. The xanthate discharge reaction, given by equation (6-5), is believed to account for this peak, and the total charge passed was found to correspond to monolayer coverage. The increase in current density at higher potential is due to the formation of dixanthogen at the surface of the galena electrode. Alternatively, xanthate ion might chemisorb with no transfer of charge by replacing an ion of similar charge at the surface. Hydroxyl ion could be adsorbed under most conditions in moderately basic medium and could be replaced readily by xanthate ion. In oxygen-deficient systems, only monolayer coverage of xanthate occurs on galena (Figure 6-1). In systems open to air, multilayer coverage occurs, and these phenomena are postulated to result from the following sequence of events.

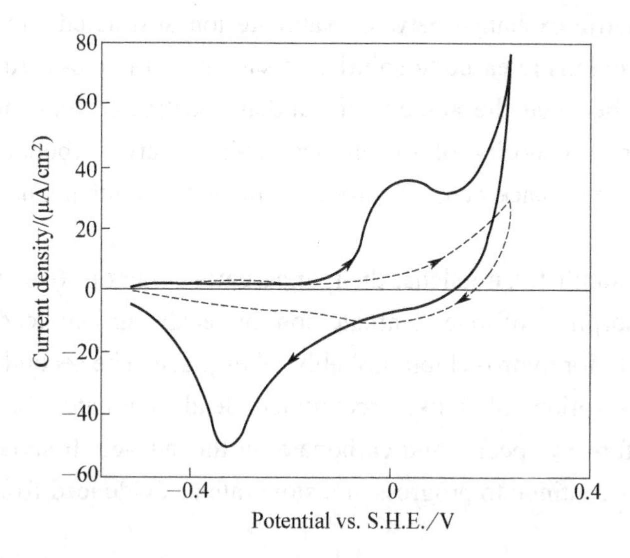

Figure 6-10 Current-potential curves for a galena electrode in 0.1 mol/L borate solution

Triangular potential sweep at 10mV/s; ethyl xanthate concentration, 0, -9.5×10^{-3} mol/L

(1) Oxidation of surface sulfide to thiosulfate and sulfate. In the presence of oxygen, galena is oxidized to lead sulfate according to equation (6-6).

$$PbS_{(s)} + 2O_{2(g)} \rightleftharpoons PbSO_{4(s)} \qquad K=10^{126} \tag{6-6}$$

With the following equilibrium expression.

$$\frac{\alpha_{PbSO_4}}{\left(\alpha_{PbS}\right)\left(P_{O_2}\right)^2} = 10^{126}$$

At equilibrium, $P_{O_2}=10^{-63}$ atm (1 atm=101325 Pa). Since the oxygen tension in air is 0.2 atm, oxidation of surface sulfide to thiosulfate and sulfate occurs spontaneously in galena systems open to the air.

(2) Metathetic replacement of surface thiosulfate and sulfate by carbonate. Since the CO_2 content of air is 300×10^{-6} by volume, with systems open to the air, carbonate ion will be present in solution, and the galena surface will carbonate at the expense of thiosulfate and sulfate. That is,

$$PbSO_{4(s)} + CO_3^{2-} \rightleftharpoons PbCO_{3(s)} + SO_4^{2-} \tag{6-7}$$

(3) Metathetic replacement of surface lead carbonate, sulfate, and thiosulfate by xanthate. At the usual flotation pH of 8 to 9, lead xanthates are more stable than lead carbonate, sulfate or thiosulfate, and lead xanthate will form by metathetic replacement of these lead salts.

$$PbCO_{3(s)} + 2X^- \rightleftharpoons PbX_{2(s)} + CO_3^{2-} \tag{6-8}$$

$$PbSO_{4(s)} + 2X^- \rightleftharpoons PbX_{2(s)} + SO_4^{2-} \tag{6-9}$$

The stoichiometric exchange between xanthate ion abstracted and reduced sulfur-oxy, sulfate and carbonate ions released to solution is shown in Table 6-4. As can be noted, there is good agreement between the amount of xanthate abstracted and the amount of anions exchanged. Further, a majority of the anions replaced are carbonate. Further evidence supporting metathetic replacement of surface anions by xanthate has been provided by utilizing calorimetry.

Adsorption of xanthate on galena, then, apparently occurs in two stages, the first stage comprising chemisorption of one xanthate ion on each surface lead ion during which exchange of xanthate for hydroxyl ion probably takes place. The second stage comprises the formation and adsorption of bulk precipitated lead xanthate formed by metathetic replacement of sulfur-oxy species and carbonate on the surface. It should be noted that the metathetic reactions continue to progress at a slow rate as evidenced from the data presented in Table 6-4.

Table 6-4 Ion exchange between galena and xanthate①

Original xanthate solution②	Xanthate ion abstracted	Ions emitted			
		Reduced sulfur-oxygen	Sulfate ions	Carbonate ions	Total ions emitted
200.0	39.0③	8.0	5.8	26.0	39.0
200.0	52.3④	12.7	12.8	—	—
200.0	58.3⑤	13.8	16.1	27.2	57.5
200.0	64.8⑥	14.1	18.7	25.8	57.5

① Added as potassium ethyl xanthate; all ions in this table expressed as milligrams potassium ethyl xanthate per liter or the stoichiometric equivalent.

② 300 mL solution added for each 30 g galena.

③ Without exposure after grinding and before xanthating.

④ Exposure 1 week.

⑤ Exposure 2 week.

⑥ Exposure 3 week.

In view of the role that bulk precipitates assume in these systems, one of the roles of the hydrocarbon chain of the collector may be that of a lower collector concentration requirement for precipitation of the collector salt, the stability of the various lead xanthates increases as the chain length increases. For example, with the same concentration of Pb^{2+}, a smaller concentration of amyl xanthate is required to form a precipitate of lead amyl xanthate than is required to form lead ethyl xanthate with ethyl xanthate. Also, of course, the longer the hydrocarbon chain, the greater is the hydrophobicity imparted to the mineral surface. Flotation recovery as a function of concentration for various xanthates is presented in Figure 6-11.

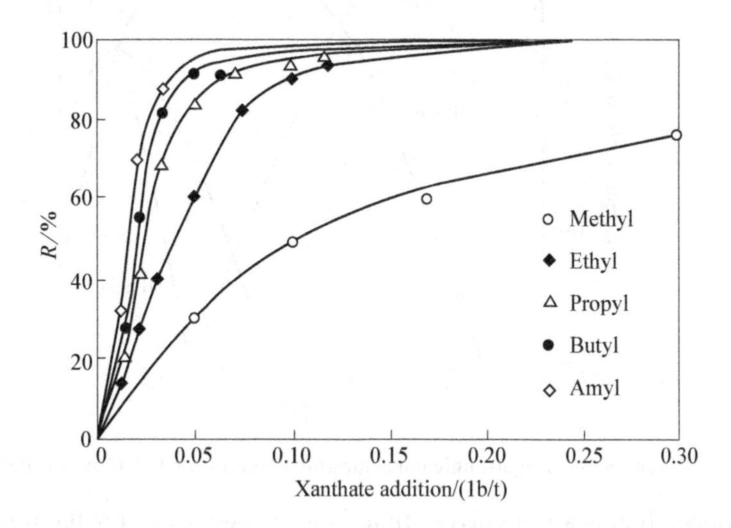

Figure 6-11 Flotation recovery of galena as a function of xanthate addition

Conditions: 100-600 mesh (100 mesh =0.15 mm, 600 mesh=0.025 mm), 0.1 lb/t terpineol and 1.0 lb/t soda ash

6.2.2 Depression of galena flotation

(1) Depression with hydroxyl. Galena depression at pH values above about pH 11 is due to the formation and stability of the lead hydroxyl species plumbite, $Pb(OH)_3^-$, on the surface relative to lead xanthate. The hydrolysis scheme of lead ion is:

$$Pb^{2+} + H_2O \rightleftharpoons PbOH^+ + H^+ \qquad K = 6.67\times10^{-7} \qquad (6\text{-}10)$$

$$PbOH^+ + H_2O \rightleftharpoons Pb(OH)_{2(aq)} + H^+ \qquad K = 1.26\times10^{-11} \qquad (6\text{-}11)$$

$$Pb(OH)_{2(aq)} + H_2O \rightleftharpoons Pb(OH)_3^- + H^+ \qquad K = 1.26\times10^{-11} \qquad (6\text{-}12)$$

The relative concentrations of these species as a function of pH when 1×10^{-4} mol/L $PbCl_2$ is added to water are presented in the distribution diagram in Figure 6-12. Plumbite in this figure is represented as $HPbO_2^-$ which is simply the unhydrated form of $Pb(OH)_3^-$. Above about pH 11, plumbite is the stable species of lead ion, and its formation on the galena surface in preference to lead xanthate is responsible for flotation depression at high values of pH.

(2) Depression with sulfide. Lead sulfide is a very insoluble compound so that

additions of sodium sulfide will result in the formation of lead sulfide rather than lead xanthate if the two aqueous species are present at the same time. If xanthate has adsorbed, a reaction of the following type will occur:

$$Pb(EX)_{2(aq)} + S^{2-} \rightleftharpoons PbS_{(s)} + 2EX^{-} \tag{6-13}$$

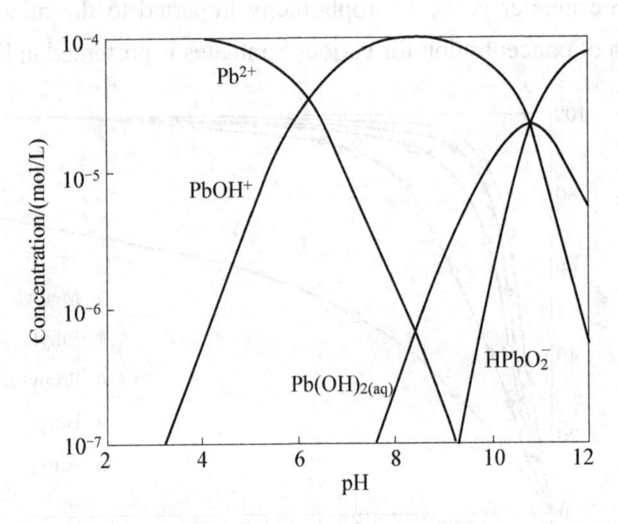

Figure 6-12 Logarithmic concentration diagram for 1×10^{-4} mol/L Pb^{2+}

Since valence bonds are involved, time lapse is necessary for the replacement of ethyl xanthate by sulfide ion.

A critical contact curve for galena as a function of sodium sulfide concentration and pH is given in Figure 6-13. Of all the sulfides listed, galena exhibits the greatest sensitivity to sulfide ion. It is effective depressant for galena.

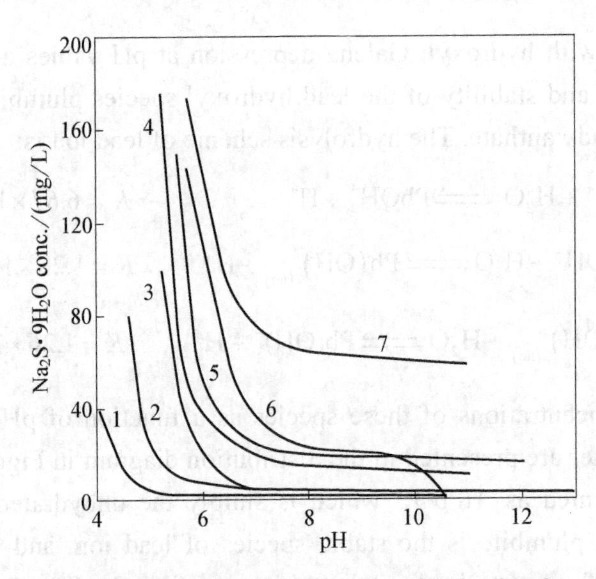

Figure 6-13 Contact curves for several sulfide minerals

Potassium ethyl xanthate concentration, 25 mg/L;

1—galena; 2—activated sphalerite; 3—chalcopyrite; 4—bornite; 5—covellite; 6—pyrite; 7—chalcocite

(3) Depression with chromate . Lead chromate is a sparingly soluble compound. As a result, when chromate salts are added to a galena system, the formation of lead chromate on the galena surface will occur. In this regard multilayers of lead chromate on galena after contact with a chromate solution was observed.

It has shown that the extent of adsorption of xanthate does not decrease when lead chromate forms on a xanthate surface. Depression is ascribed to the strong hydration of chromate adsorbed on the surface whereby the hydrophobicity of the collector coating is overcome.

(4) Depression with cyanide. With respect to depression of galena in the presence of sodium cyanide, cyanide ion does not complex lead ion at moderate concentration. As a result, cyanide concentration has no effect on bubble contact on galena, and the response shown in Figure 6-14 is due to hydroxyl ion and not cyanide.

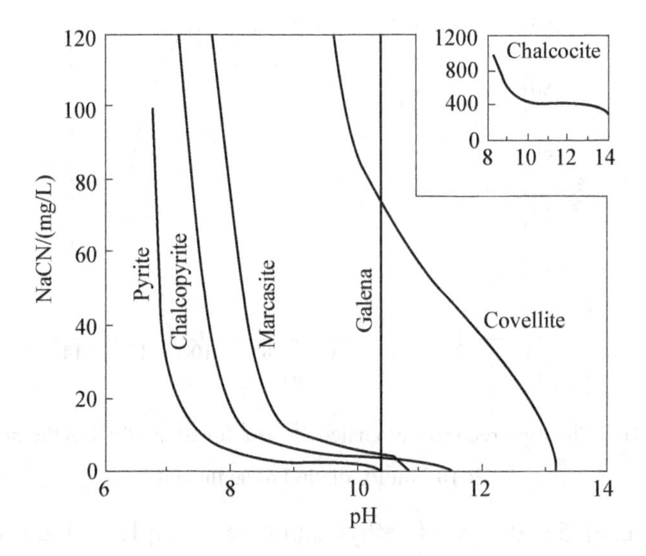

Figure 6-14 Contact curves for several sulfide minerals

6.3 Flotation Chemistry of Chalcocite

6.3.1 Chalcocite flotation

Chalcocite (Cu_2S) and chalcopyrite ($CuFeS_2$) are the two most commonly floated copper sulfide minerals. Bornite (Cu_5FeS_4), covellite (CuS) and enargite (Cu_3AsS_4) are normally present in lesser quantities.

Both chalcocite and chalcopyrite are floated readily with common sulfhydryl collectors. The response of chalcocite to flotation with various collectors is presented in Figures 6-15 and 6-16. Complete flotation is effected from about pH 2 to over pH 13 with dithiophosphatogen, the dimer of dithiophosphate.

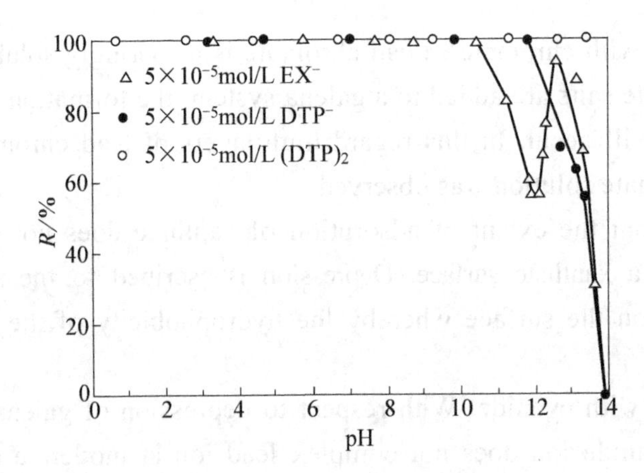

Figure 6-15 Flotation recovery of chalcocite as a function of pH with various additions of ethyl xanthate (EX^-), diethyl dithiophosphate (DTP^-) and diethyl dithiophosphatogen[$(DTP)_2$]

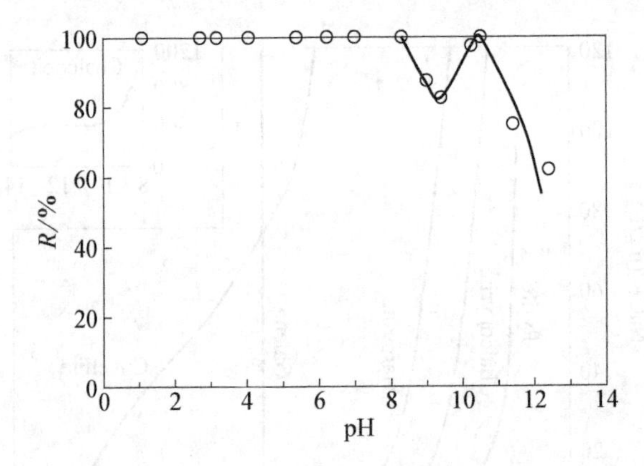

Figure 6-16 Flotation recovery of chalcocite as a function of pH in the presence of 2×10^{-5} mol/L diethyl dixanthogen

In the presence of 5×10^{-5} mol/L ethyl xanthate, complete flotation is possible until about pH 10, above which a slight region of depression occurs which is followed by an increase in flotation at higher values of pH. A similar region of depression exists when chalcocite is floated with diethyl dixanthogen. Comparison of these data with the stability regions for Cu_2S, $CuEX_{(s)}$, $(EX)_{2(aq)}$ and with the Eh values measured in this system, shows that at the pH values at which depression occurs, the stable species are $Cu(OH)_{2(s)}$ and $(EX)_{2(aq)}$. See Figure 6-17. These facts indicate that depression is due to the formation of $Cu(OH)_{2(s)}$ in preference to $CuEX_{(s)}$.

The active species of collector when xanthate is added to the chalcocite system is xanthate ion. Dixanthogen does not form on the chalcocite surface. See Table 6-1.

Similarly to galena, xanthate adsorption on chalcocite is also a two stage process. The presence of an unleachable xanthate species on the chalcocite surface following xanthate adsorption has been demonstrated. These authors also demonstrated that following the formation of an unleachable chemisorbed layer, multilayers of cuprous xanthate form and adsorb on the surface. The relative distribution between unleachable amyl xanthate, cuprous amyl xanthate, and xanthate remaining in solution is presented as a function of xanthate addition in Figure 6-18. With relatively high levels of addition of xanthate, e.g. 6 meq, approximately 25 percent of the xanthate added is unleachable, while the remaining 75 percent is present as cuprous xanthate on the chalcocite surface.

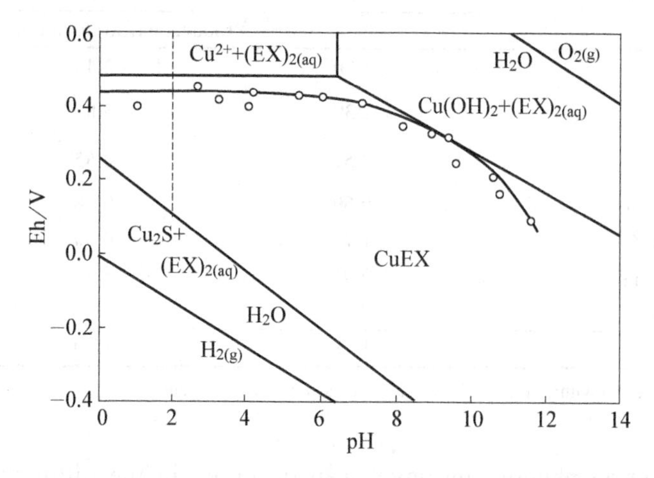

Figure 6-17 Stability relations of Cu_2S, CuEX, $Cu(OH)_2$ and $(EX)_{2(aq)}$

Assumed activities: dissolved sulfur=1×10^{-4} mol/L

$Cu^{2+}=2\times10^{-4}$ mol/L; $(EX)_{2(aq)}=1.3\times10^{-5}$mol/L; measured Eh values—open circles

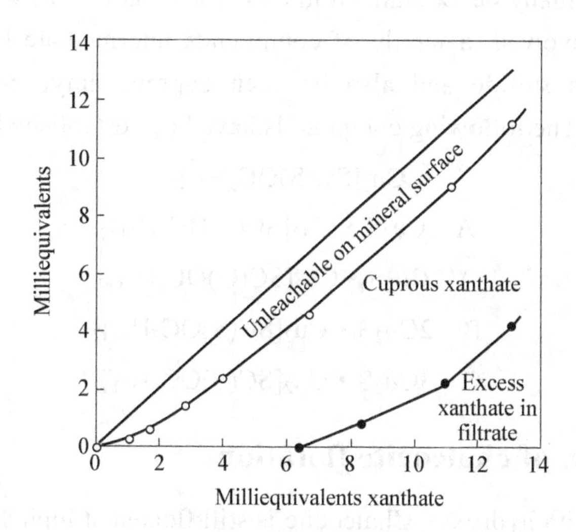

Figure 6-18 Distribution of amyl xanthate species after adsorption of amyl xanthate on chalcocite

Ion exchange experiments similar to those in the galena-amyl xanthate system were also conducted in the chalcocite-amyl xanthate system. These results are presented in Table 6-5 and show that the two principal anions exchanged when xanthate chemisorbs on chalcocite are carbonate and hydroxyl. The equivalence between xanthate abstracted and anions released is not as close as that observed in the galena-xanthate system.

Table 6-5 Effect of time of grinding on products from chalcocite

Substance	Products leached from surface/mg		
	Grinding 1.25h	Grinding 2.5h	Grinding 5h
Cuprous xanthate	1670	1524	1224
Compounds A, A′, and B	0	171	246
Compound C	0	0	177
	Products in filtrate/mmol		
OH^-	5.30	3.43	3.48
CO_3^{2-}	2.39	3.86	3.95
SO_4^{2-}	0.51	0.65	0.93
Reducing ions	0.00	0.18	0.16
Residual X^-	1.77	0.30	0.00
Total anions	9.97	8.41	8.52
Potassium cation	9.73	8.49	6.63
pH	11.8	12.0	12.0

Note: samples (50 g) treated with 10 millimoles (2.02 g) potassium-amyl xanthate.

Similar adsorption phenomena have been observed in the dithiophosphate-chalcocite system. Chemisorption of dithiophosphate as well as the formation of cuprous and cupric diethyl dithiophosphate on the surface of chalcocite was observed.

Equilibrium is actually never attained in this system as well as with galena. When long time of grinding is involved, a family of compounds intermediate between cuprous amyl xanthate and cuprous sulfide and also between cuprous amyl monothiocarbonate and cuprous sulfide form. The following compounds have been established:

$$Cu_2[SC(S)OC_5H_{11}]_2$$

$$A \quad Cu_2S \cdot Cu_2[SC(S)OC_5H_{11}]_2$$

$$A' \quad Cu_2S \cdot Cu_2[SC(O)OC_5H_{11}]_2$$

$$B \quad 2Cu_2S \cdot Cu_2[SC(S)OC_5H_{11}]_2$$

$$C \quad 3Cu_2S \cdot Cu_2[SC(S)OC_5H_{11}]_2$$

6.3.2 Depression of chalcocite flotation

(1) Depression with hydroxyl. Chalcocite is still floated at high values of pH due to the stability of cuprous ethyl xanthate relative to cuprous hydroxide, their K_{sp} being 5.2×10^{-20} and 2×10^{-15}, respectively. Depression occurs at about pH 14 in the presence of this collector, and calculations show that cuprous hydroxide becomes stable with respect to cuprous ethyl xanthate at about this pH. See also the Eh-pH diagram for the Cu/H_2O system and the measured Eh values in Figure 6-17.

(2) Depression with sulfide. Sulfide ion should function in a manner similar to that in the case of galena. Care should be exercised, however, when comparing the sulfide ion

concentration necessary for depression or to preclude air bubble contact because oxidation potentials (Eh) were not considered when these measurements were made (Figure 6-13). With changes in sulfide concentration, Eh will have changed, and the stability regions of metal sulfides will have changed.

(3) Depression with cyanide. In the presence of cyanide and absence of iron, chalcocite is depressed only with very high additions of cyanide (Figure 6-14). This phenomenon is probably due to the stability of cuprous ethyl xanthate relative to the cuprocyanide complex.

6.4 Flotation Chemistry of Sphalerite

6.4.1 Sphalerite flotation

The flotation characteristics of sphalerite have received considerable attention both in the absence and presence of activating ions. Some investigators have observed flotation with ethyl and amyl xanthates in the absence of activators, while others have not. It seems likely that these differences in response may have been due to differences in the oxidation characteristics of the sphalerites involved. That is, the formation of bulk precipitates of zinc xanthates on sphalerite has been shown to be necessary for flotation in the absence of activators. With those sphalerites that are refractory to oxidation, only a limited quantity of Zn^{2+} will be available for the formation of multilayers of zinc xanthate on the surface.

In this regard xanthate adsorption on sphalerite is similar to that on chalcocite and galena in which xanthate appears to adsorb via two stages. The first stage involves chemisorption of an initial layer of xanthate at 1:1 coordination, while the second apparently involves the formation and adsorption of bulk precipitated zinc xanthate on the sphalerite surface.

After exposure to xanthate, the collector species on the surface readily identifiable with infrared analysis is bulk precipitated zinc xanthate. The major absorption bonds of zinc hexyl xanthate occur at 1375, 1227, 1112, and 1047 cm^{-1}. After contact with hexyl xanthate at pH 5.5, absorption bonds are measured at 1380, 1227, 1112, and 1047 cm^{-1} on sphalerite. These results are similar to those presented with sphalerite after contact with dodecyl xanthate. Dixanthogen is not present on the surface.

These observations are in agreement with those of other investigators. It concluded that two forms of adsorption occur in this system. Weakly attached xanthate is removed by water washing, while firmly attached xanthate is dissolved with pyridine. It shows that under flotation conditions, four to five times monolayer coverage is adsorbed.

Additional evidence for the presence of a separate phase on the sphalerite surface after contact with xanthate is provided by the electrokinetic data shown in Figure 6-19. The values of the zeta potential of sphalerite contacted with hexyl xanthate are between those of sphalerite and zinc hexyl xanthate. Further, the zeta potential of the precipitate of zinc xanthate reverses sign at approximately the pH value at which zeta reversal of sphalerite in contact with xanthate occurs.

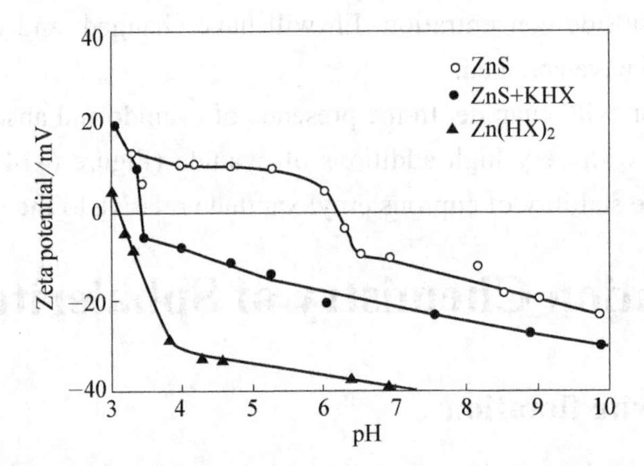

Figure 6-19 Zeta potential of zinc hexyl xanthate and sphalerite in the absence and presence of 1.25×10^{-4}mol/L hexyl xanthate as a function of pH

Flotation response of sphalerite with amyl xanthate is presented in Figure 6-20. Excellent flotation is obtained at pH 3.5, while modest recovery is obtained even at pH 9. Significantly also is the fact that there is no difference in flotation response between sphalerite and marmatite.

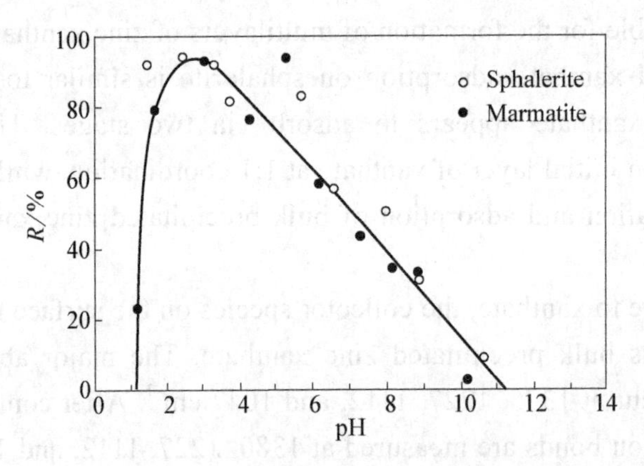

Figure 6-20 Flotation recovery of sphalerite and marmatite as a function of pH with 2.5×10^{-4} mol/L amyl xanthate

Flotation recovery of sphalerite is shown as a function of xanthate concentration and hydrocarbon chain length at constant pH in Figure 6-21. As the carbon content of the hydrocarbon chain is reduced, the concentrations necessary for flotation are increased. In the case of ethyl xanthate, a concentration of about 2×10^{-2} mol/L is necessary for complete flotation whereas a concentration of about 3×10^{-4} mol/L amyl xanthate results in the same response. Correlation of these flotation results with the solubility products of the corresponding zinc alkyl xanthates indicates the necessity for the formation and adsorption of multilayers of zinc xanthate having a stoichiometry of 1:2.

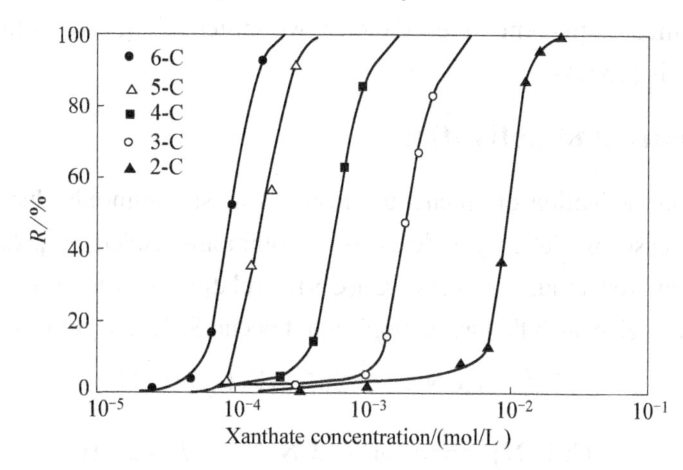

Figure 6-21 Flotation recovery of sphalerite as a function of xanthate concentration and hydrocarbon chain length at pH 3.5

6.4.2 Activation

Advantage is taken in practice of the fact that sphalerite does not float in alkaline medium with modest levels of ethyl xanthate. Prevention of activation is ensured by suitable reagent schedules, and other sulfides such as chalcocite, chalcopyrite and galena are floated selectively from sphalerite.

When it is desired to float sphalerite, activation is accomplished by adding a metal ion whose metal sulfide is more stable than ZnS. A number of metal ions possess this property, notably cuprous, cupric, mercurous, mercuric, silver, lead, cadmium and antimony.

The most commonly added activator is copper sulfate. In a study utilizing radioactive copper, it showed that Cu^{2+} displaces Zn^{2+} from the sphalerite lattice. Exchange is quite rapid until three layers of zinc are replaced. Additional exchange follows a parabolic rate law typical of diffusion-controlled systems.

The activation reaction with Cu^{2+} is represented by equation (6-14).

$$ZnS_{(s)} + Cu^{2+}_{(aq)} \longrightarrow CuS_{(s)} + Zn^{2+}_{(aq)} \qquad K = 9\times 10^{10} \tag{6-14}$$

Cupric ion will replace zinc until the activity of Zn^{2+} is 9×10^{10} times that of Cu^{2+} in solution. Following activation with Cu^{2+}, the flotation response of sphalerite is similar to copper sulfide minerals.

Other activation reactions are:

$$ZnS_{(s)} + 2Ag^{+} \rightleftharpoons Ag_2S_{(s)} + Zn^{2+} \qquad K = 9\times 10^{26} \tag{6-15}$$

$$ZnS_{(s)} + Pb^{2+} \rightleftharpoons PbS_{(s)} + Zn^{2+} \qquad K = 9\times 10^{3} \tag{6-16}$$

It showed that the abstraction of Ag^{+} by sphalerite is rapid and that the sphalerite sample turned black in a short period of time. These authors also showed that the uptake of silver occurs indefinitely and that after two months of reaction, a plus 325-mesh sample of sphalerite contained 18 percent silver. Exactly two moles of Ag^+ are exchanged for one mole of Zn^{2+} during this process.

6.4.3 Prevention of activation

Unintentional activation of sphalerite in ores is most commonly due to Cu^{2+} and Pb^{2+} in solution. In the case of Cu^{2+}, cyanide is most commonly added to prevent activation. The stability of the cuprocyanide complex, $Cu(CN)_2^-$, relative to $Zn(CN)_4^{2-}$ results in ratios of dissolved copper to zinc such that activation cannot occur. Relevant equilibria are as follows:

$$Cu^{2+} + HCN \rightleftharpoons Cu^+ + H^+ + \frac{1}{2}(C_2N_2)_{(g)} \tag{6-17}$$

$$Cu(CN)_2^- \rightleftharpoons Cu^+ + 2CN^- \qquad K = 2\times10^{-24} \tag{6-18}$$

$$Zn(CN)_2^- \rightleftharpoons Zn^+ + 2CN^- \qquad K = 2\times10^{-18} \tag{6-19}$$

$$HCN_{(aq)} \rightleftharpoons H^+ + CN^- \qquad K = 4\times10^{-10} \tag{6-20}$$

In the case of Pb^{2+}, activation can be prevented when the activity of Zn^{2+} is 1000 times that of Pb^{2+} in solution. Since sphalerite commonly resists oxidation, very little Zn^{2+} dissolves from this mineral. Zinc sulfate is therefore added, and by virtue of the equilibria involving basic lead carbonate and zinc hydroxide, the activity ratio of Zn^{2+}/Pb^{2+} is higher than the equilibrium ratio, and activation cannot occur. The equilibria involving the lead and zinc species are:

$$Pb_3(CO_3)_2(OH)_{2(S)} \rightleftharpoons 2PbOH^+ + Pb^{2+} + 2CO_3^- \qquad K = 3.16\times10^{-32} \tag{6-21}$$

$$PbCO_{3(S)} \rightleftharpoons Pb^{2+} + CO_3^{2-} \qquad K = 1.5\times10^{-13} \tag{6-22}$$

$$Pb^{2+} + H_2O \rightleftharpoons PbOH^+ + H^+ \qquad K = 6.67\times10^{-7} \tag{6-23}$$

$$Zn(OH)_{2(S)} \rightleftharpoons Zn^{2+} + 2OH^- \qquad K = 4.5\times10^{-17} \tag{6-24}$$

With an externally fixed pressure of CO_2, namely, $P_{CO_2} = 3\times10^{-4}$ atm which is that in air, the following equilibria apply.

$$CO_{2(g)} \rightleftharpoons CO_{2(aq)} \qquad K = 3.6\times10^{-2} \tag{6-25}$$

$$CO_{2(aq)} + H_2O \rightleftharpoons HCO_3^- + H^+ \qquad K = 3.98\times10^{-2} \tag{6-26}$$

$$HCO_3^- \rightleftharpoons CO_3^{2-} + H^+ \qquad K = 5.02\times10^{-11} \tag{6-27}$$

From equation (6-25).

$$[CO_2]_{(aq)} = 1.08\times10^{-5} \tag{6-28}$$

Substituting into equation (6-26).

$$[HCO_3^-] = 4.3\times10^{-12}/[H^+] \tag{6-29}$$

and from equation (6-27).

$$\left[CO_3^{2-}\right]=2.16\times10^{-22}/\left[H^+\right]^2 \tag{6-30}$$

Combining equations (6-21), (6-23) and (6-30) yields $[Pb^{2+}]=1.15\times10^{8}/[H^+]^2$. At pH 9, for example, the Pb^{2+} activity is 1.15×10^{-10}. From the solubility product of $Zn(OH)_{2(S)}$, equation (6-24), the Zn^{2+} activity at pH 9 is 4.5×10^{-7}. The activity ratio of Zn^{2+}/Pb^{2+} is $4.5\times10^{-7}/1.15\times10^{-10}$, which equals 3.91×10^{3}, and which is greater than the equilibrium constant for the activation reaction, equation (6-16). As a result the activation reaction cannot occur under these conditions.

In addition to these thermodynamic considerations, there is considerable evidence which indicates that the colloids of zinc salts, formed under conditions in which precipitation occurs, function as depressants for sphalerite. These include precipitates of zinc hydroxide, zinc carbonate, zinc sulfite, and zinc cyanide.

The depressant role of zinc hydroxide colloids was first presented by Malinovsky. His observations were later confirmed by Livshitz and Idelson, who also demonstrated that the extent of depression of sphalerite and the concentration of colloidal zinc hydroxide occurring in the pulp are directly related.

Debrivnaya has shown that copper-activated sphalerite is effectively depressed in solutions of $ZnSO_4/Na_2CO_3$ of pH values greater than about pH 6, while the floatabilities of chalcopyrite and pyrite are hardly affected.

Grossman and Khadshiev confirmed this finding. Their results showed that both sphalerite and chalcopyrite take up the equivalent of many tens of monolayers of zinc colloids. The chalcopyrite retains its hydrophobic character under these conditions, while the sphalerite is hydrophilic. As the adhering colloids are successively removed from the sphalerite, floatability of the sphalerite is increased until the equivalent of three monolayers remains, and complete flotation is obtained. These authors concluded that basic zinc carbonate was formed under these conditions of sphalerite depression.

Two possible explanations can be suggested for the adsorption of these colloidal precipitates on the sphalerite surface:

① The colloidal precipitate is formed in the bulk solution and attaches itself by physical forces to the surface, i.e. agglomeration or coagulation.

② The colloidal precipitate nucleates and grows at surface sites.

Additional work is obviously required before an understanding of these phenomena will be achieved.

6.5 Flotation Chemistry of Pyrite

6.5.1 Pyrite flotation

The mechanisms by which pyrite is floated with sulfhydryl collectors are well understood. The species of xanthate responsible for flotation in the presence of short-chained xanthates is dixanthogen. This conclusion in the presence of short-chained xanthates is dixanthogen. This conclusion has been drawn from spectroscopic, electrokinetic, electrochemical, thermochemical and flotation data.

The presence of dixanthogen only on the pyrite surface after contact with xanthate has been shown clearly with infrared spectroscopy. The principal absorption bands of diamyl dixanthogen occur at 1021 and 1258 cm^{-1}. After contact with amyl xanthate, the principal absorption bands of pyrite occur at 1028 and 1258 cm^{-1} which correspond closely with those of dixanthogen.

The adsorption of an electrically-neutral species on the surface is also apparent from the data presented in Figure 6-22. That is, the zeta potential of pyrite is the same in the absence and presence of ethyl xanthate. This should be contrasted with the electrokinetic behavior of pyrite conditioned with oleate. Chemisorption of oleate occurs on the surface with a resultant change in zeta potential (Figure 6-23).

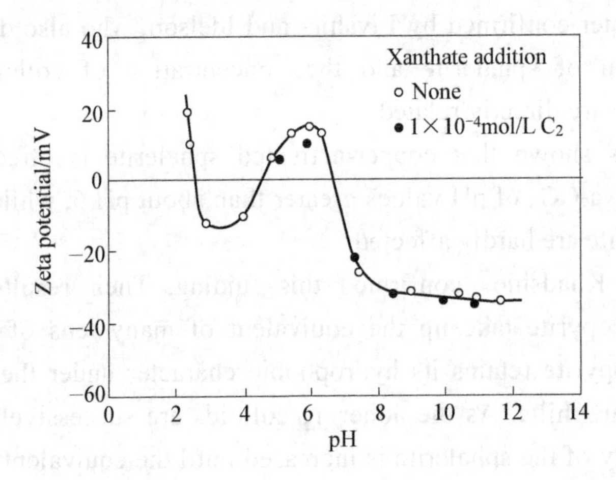

Figure 6-22 Zeta potential of pyrite as a function of pH in the absence and presence of ethyl xanthate in the presence of air

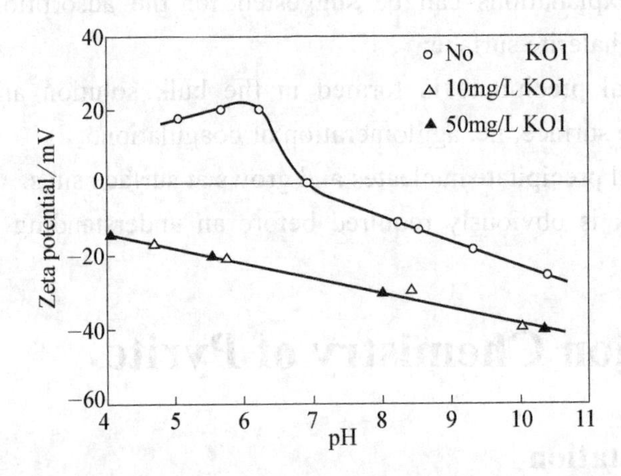

Figure 6-23 Zeta potential of pyrite as a function of pH in the absence and presence of potassium oleate

Dixanthogen forms by anodic oxidation of xanthate ion on the surface of pyrite coupled with cathodic reduction of adsorbed oxygen. That is:

$$2X^- \rightleftharpoons X_2 + 2e^- \qquad \text{Anodic} \tag{6-31}$$

$$\left(\tfrac{1}{2}\right)O_{2(ads)} + H_2O + 2e^- \rightleftharpoons 2OH^- \qquad \text{Cathodic} \tag{6-32}$$

Where X^- represents xanthate ion and X_2 represents dixanthogen.

Since sulfides are electronic conductors, electron transfer occurs through the solid. Schematically,

$$2e^- + \frac{1}{2}O_2 + H_2O \rightarrow 2OH^-$$

FeS_2 $2e^-$

$$2e^- + X_2 \rightarrow 2X^-$$

The overall reaction is,

$$2X^- + \frac{1}{2}O_{2(ads)} + H_2O \rightleftharpoons X_2 + 2OH^- \tag{6-33}$$

The adsorption reaction kinetics can be explained from electrochemical theory. This reaction occurs up to about pH 11, above this pH, xanthate ion is the stable species of xanthate.

Flotation of pyrite, then, is possible below pH 11 with short-chain xanthates but is depressed above about pH 11. These phenomena are shown clearly in Figure 6-24. With low levels of addition of ethyl xanthate, two regions of flotation are obtained from about pH 3 to pH 9. The intermediate region of depression is not related to a lack of dixanthogen, however.

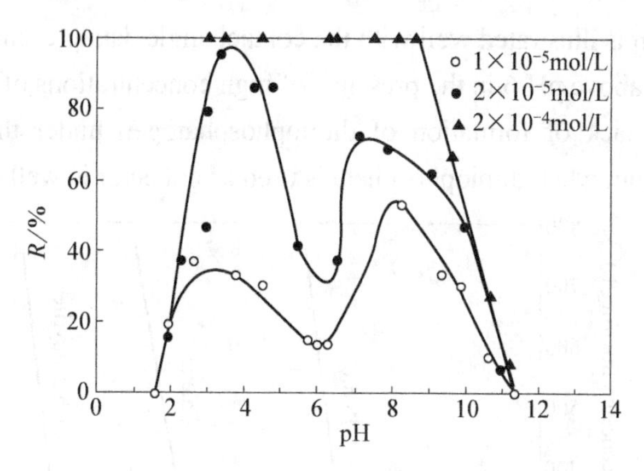

Figure 6-24 Flotation recovery of pyrite as a function of pH with various additions of potassium ethyl xanthate

Another collector has been shown to function similarly in the pyrite system, and this is dithiophosphate. Dithiophosphate is more difficult to oxidize to its dimer, dithiophosphatogen, however, than xanthate is to dixanthogen. That is,

$$X_{2(L)} + 2e^- \rightleftharpoons 2X^- \qquad E^0 = -0.06\text{V} \tag{6-34}$$

$$(DTP)_{2(L)} + 2e^- \rightleftharpoons 2DTP^- \qquad E^0 = -0.25\text{V} \tag{6-35}$$

Where X^- and DTP^- represent xanthate and dithiophosphate, respectively. The structural formulas of dixanthogen, X_2, and dithiophosphatogen, $(DTP)_2$, are:

Dixanthogen

Dithiophosphatogen

Where R represents the hydrocarbon chain

Diethyl dithiophosphatogen has been found experimentally to form at pH 4 and below in the presence of pyrite but not at pH 6 and above. See Table 6-6.

Table 6-6 Diethyl dithiophosphate oxidation in the absence and presence of 100-200 mesh (200 mesh=0.075 mm)-pyrite (reaction time: 10 min)

pH	Pyrite/g	DTP^-/(mol/L)	Fe^{3+}/(mol/L)	$(DTP)_2$ formed /(mol/L)
2.0	0	5×10^{-4}	2.5×10^{-4}	7×10^{-6}
3.0	0	5×10^{-4}	2.5×10^{-4}	nil
2.0	2	5×10^{-4}	0	4.3×10^{-5}
3.0	2	5×10^{-4}	0	3.3×10^{-5}
3.0	2	5×10^{-4}	0	$<7\times10^{-6}$
4.0	2	5×10^{-4}	0	$<7\times10^{-6}$
6.0	2	5×10^{-4}	0	nil
8.0	2	5×10^{-4}	0	nil

This phenomenon is illustrated well with the contact angle data presented in Figure 6-25. The lack of bubble contact above pH 6 in the presence of high concentrations of dithiophosphate is due to the instability and lack of formation of dithiophosphatogen under these conditions. Pyrite rejection in basic medium when dithiophosphate is used as collector is well known.

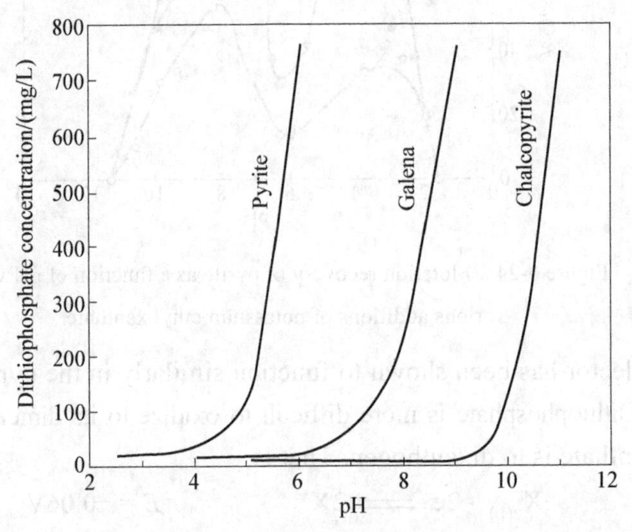

Figure 6-25 Contact curves for several sulfide minerals as a function of diethyl dithiophosphate concentration and pH

6.5.2 Depression of pyrite flotation

(1) Depression with hydroxyl. Hydroxyl ion functions as a depressant for a number of reasons. First, with relatively high hydroxyl activities, e.g. those above pH 11, xanthate oxidation to dixanthogen does not occur. The reaction in equation (6-33) proceeds to the left under these conditions. Secondly, also above about pH 11, the surface of pyrite will consist of ferric hydroxide which fact can be noted in Figure 6-26.

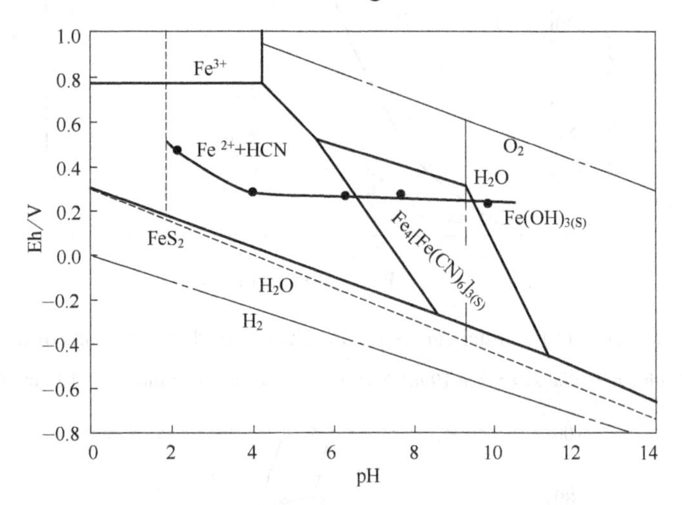

Figure 6-26 Stability of FeS_2, $Fe(OH)_3$, and $Fe_4[Fe(CN)_6]_3$ at 3×10^{-4} mol/L total dissolved sulfur, 5×10^{-5} mol/L total dissolved iron and 6×10^{-4} mol/L cyanide addition

Black circles indicate Eh values corresponding to Curve(A), Figure 6-29

Electrochemical measurements with a pyrite electrode confirm these findings. Current versus potential curves are presented in Figure 6-27 for a pyrite electrode in the presence and absence of ethyl xanthate for selected pH values. Note that in the presence of ethyl xanthate (solid curve) that the polarization curve is independent of pH and also that significant current is observed when the pyrite electrode potential exceeds the reversible Nernst potential for the $EX^-/(EX)_2$ couple, 0.08 volt. On the other hand, in the absence of ethyl xanthate (broken curves), the current observed is due to oxidation at the pyrite surface which is pH dependent. As the pH is increased, oxidation is facilitated, and a lower potential is required to achieve a given current. Finally, at pH 11.4 oxidation of pyrite occurs at a lower potential than xanthate oxidation to dixanthogen. Hence, the stability of the ethyl xanthate is preserved at the expense of the pyrite which corrodes. These observations also demonstrate that stabilization of ethyl xanthate and the formation of ferric hydroxide account for the depression of pyrite at high values of pH.

Calcium ion from lime in pH adjustment also contributes to pyrite depression. As shown in Figure 6-28, depression is achieved about one pH unit lower when CaO is used for pH adjustment as compared with KOH and K_2CO_3. With reference to Figure 6-22, one of the

pzc of pyrite in the presence of air is pH 6.8. Above pH 6.8, then, the surface of pyrite is negatively charged, and Ca^{2+} adsorbs readily by electrostatic attraction. The presence of Ca^{2+} ion obviously hinders the oxidation reaction on the pyrite surface.

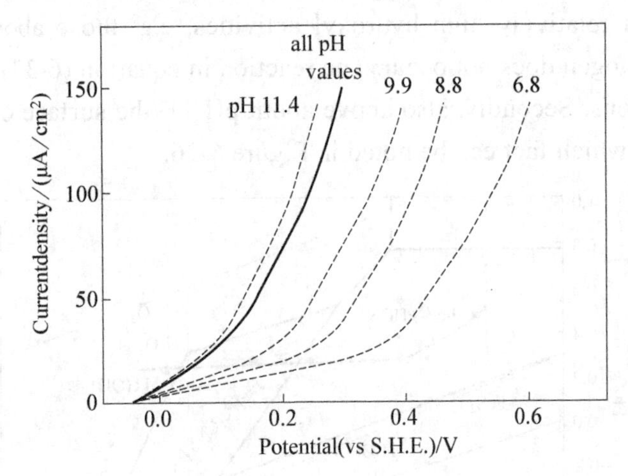

Figure 6-27 Current-potential curves for a pyrite electrode in solutions of different pH

Anodic potential sweeps at 10 mV/s;ethyl xanthate concentration:0-5×10^{-3} mol/L

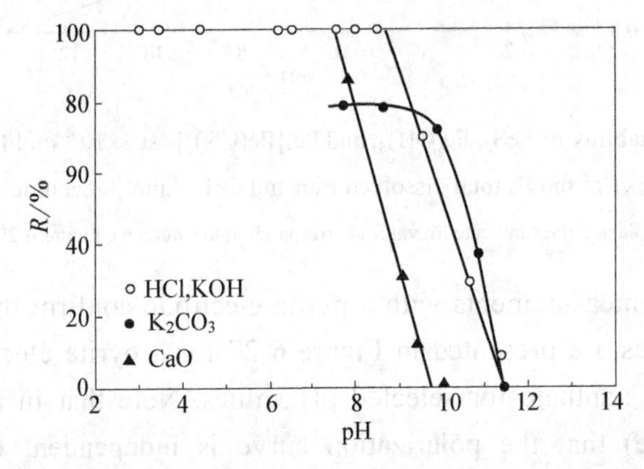

Figure 6-28 Flotation recovery of pyrite as a function of pH with 2×10^{-4} mol/L ethyl xanthate

(2) Depression with cyanide. A detailed study was undertaken by Elgillani and Fuerstenau to establish which species in the pyrite-cyanide-xanthate system is responsible for depression and then to determine the conditions under which this species will adsorb.

The flotation response of pyrite in the absence and presence of potassium cyanide with 1×10^{-4} mol/L ethyl xanthate is given in Figure 6-29. In the presence of 6×10^{-3} mol/L KCN, pyrite is depressed approximately four units in pH below that observed in the absence of cyanide.

Flotation recovery in the absence and presence of potassium ferrocyanide is presented in Figure 6-30. The fact that depression occurs at approximately the same pH conditions in the presence of 6×10^{-4} mol/L KCN and 1×10^{-4} mol/L $K_4Fe(CN)_6$ can be noted.

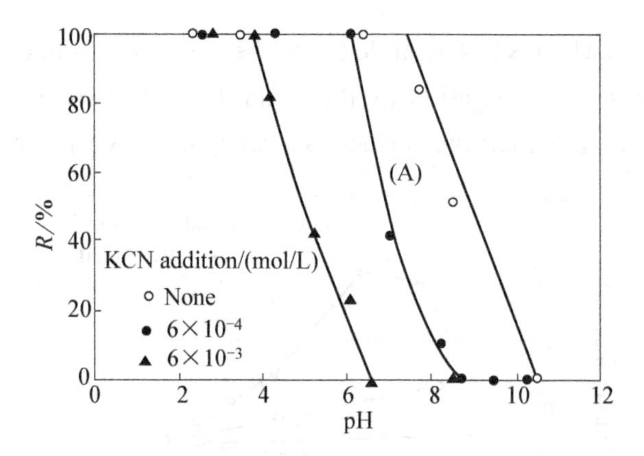

Figure 6-29 Flotation recovery of pyrite as a function of pH with 5×10^{-4} mol/L ethyl xanthate in the absence and presence of cyanide

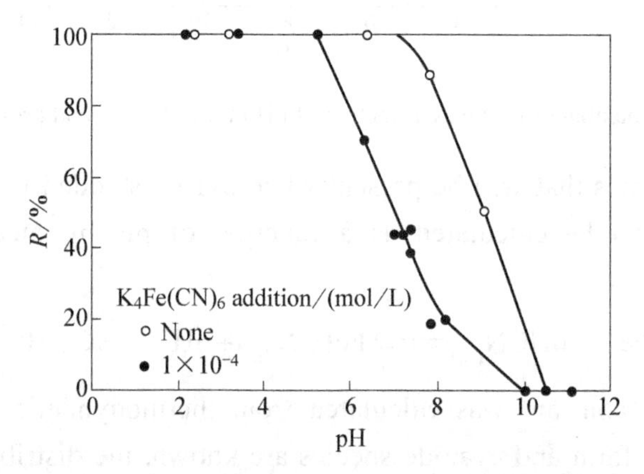

Figure 6-30 Flotation recovery of pyrite as a function of pH with 1×10^{-4} mol/L ethyl xanthate in the absence and presence of ferrocyanide

Next, electrokinetic determinations made in the presence of various addictions of KCN and $K_4Fe(CN)_6$ should be compared. See Figures 6-31 and 6-32. Chemisorption of ferrocyanide

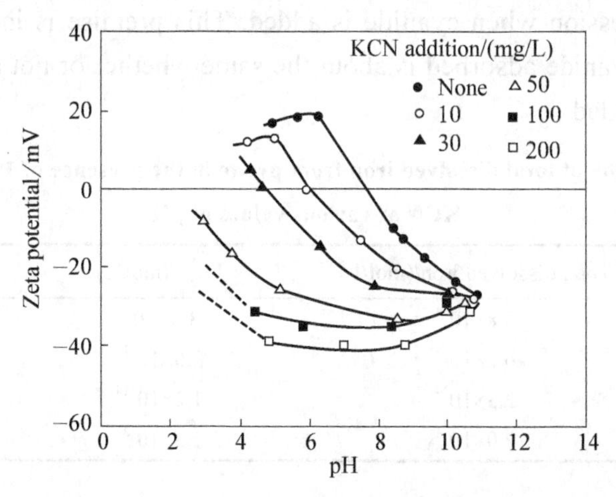

Figure 6-31 Zeta potential of pyrite as a function of pH in the absence and presence of cyanide

and a cyanide species (when KCN is added) can be noted up to about pH 11. That is, the zeta potential becomes more negative as the quantities of cyanides and ferrocyanide are increased in a pH region in which the surface is already negatively charged.

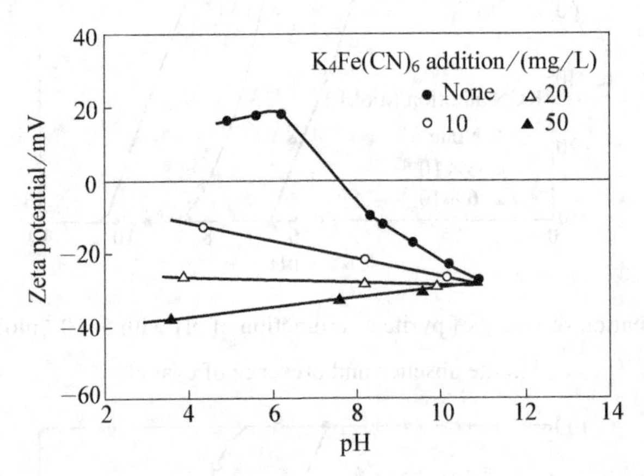

Figure 6-32 Zeta potential of pyrite as a fraction of pH in the absence and presence of ferrocyanide

The cyanide species that will be present when pyrite is added to a solution containing potassium cyanide can be calculated as a function of pH by means of the following expression:

$$Fe^{2+} + 6HCN_{(aq)} \rightleftharpoons Fe(CN)_6^{4-} + 6H^+ \qquad K = 10^{-22} \tag{6-36}$$

The equilibrium constant was calculated from thermodynamic data. When the total amounts of dissolved form and cyanide species are known, the distribution of uncompleted ferrous ion and ferrocyanide ion can be calculated (Table 6-7).

At pH 4.4 and below, almost all of the dissolved iron exists as ferrous ion. At higher pH values, this concentration is reduced until at pH 6.4 and above, almost all of the iron exists in the ferrocyanide form. As a result it seems likely that $Fe(CN)_6^{4-}$ is the active species responsible for depression when cyanide is added. This premise is in accord with the fact that the amount of cyanide adsorbed is about the same whether or not an equivalent amount of $K_4Fe(CN)_6$ is added.

Table 6-7 Distribution of total dissolved iron from pyrite in the presence of 1.54×10^{-2} mole per liter KCN at various values of pH

pH	Total dissolved iron/(mol/L)	Fe^{2+} /(mol/L)	$Fe(CN)_6^{4-}$ /(mol/L)
4.4	3.8×10^{-4}	3.8×10^{-4}	1.2×10^{-10}
6.4	6.2×10^{-5}	2.0×10^{-10}	6.2×10^{-5}
7.2	2.3×10^{-4}	1.2×10^{-14}	2.3×10^{-4}
9.0	2.0×10^{-4}	2.3×10^{-24}	2.0×10^{-4}

The formation of ferric ferrocyanide on the surface of pyrite is proposed to occur according to the following half cell:

$$7Fe^{2+} + 18HCN \rightleftharpoons Fe_4\left[Fe(CN)_6\right]_3 + 18H^+ + 4e^- \tag{6-37}$$

This equation represents the overall reaction, which must occur in several steps. The mineral must produce some ferrous ion in solution which reacts with CN^- to form $Fe(CN)_6^{4-}$. Most pyrites yield sufficient iron in solution especially after oxidation. Cyanide becomes a much more effective depressant when pyrite has been oxidized (Figure 6-33).

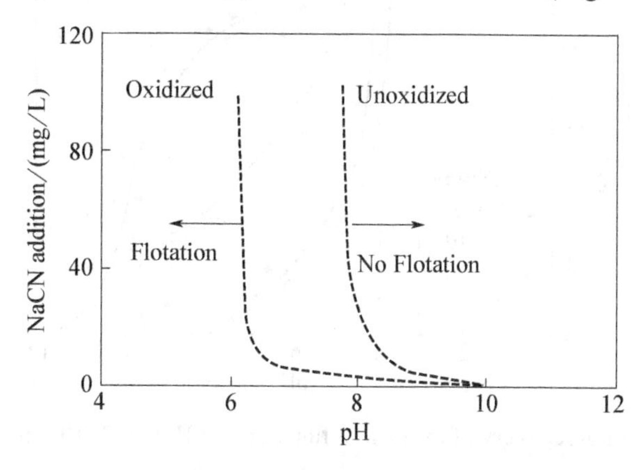

Figure 6-33 Contact curves for oxidized and unoxidized pyrite as a function of cyanide addition and pH

Potassium ethyl xanthate, 25 mg/L; 35℃

It should be mentioned that in addition to oxidation providing more iron for cyanide depression, excessive oxidation of pyrite can greatly reduce the floatability of pyrite without depressants.

The domains of stability of FeS_2, $Fe_4[Fe(CN)_6]_3$, $Fe(OH)_3$, Fe^{2+}, Fe^{3+} and HCN are presented in the Eh-pH diagram in Figure 6-26. Also presented in this figure are the Eh values measured at various values of pH. The darkened circles indicate Eh values corresponding to Curve(A), Figure 6-29. The pH value above which , $Fe_4[Fe(CN)_6]_3$ is the stable species closely approximates the pH at which depression is obtained.

Similar conclusions on the formation of a surface species of ferric ferrocyanide resulting in depression of pyrite have been drawn from an electrochemical investigation.

Note also that $Fe(OH)_3$ is the stable species above about pH 11 under these conditions. This fact can also be seen from the electrokinetic data in Figures 6-31 and 6-32. Above pH 11 the zeta potential is independent of cyanide and ferrocyanide addition.

(3) Depression with sulfite. Since dixanthogen is the species responsible for pyrite flotation, reagents which are more reducing than the xanthate-dixanthogen couple should function as depressants for pyrite. An average value of −0.06 volt (IUPAC) has been established for the standard cell potential of the dixanthogen-xanthate half cell:

$$X_{2(l)} + 2e^- \rightleftharpoons 2X^- \qquad E^0 = -0.06V \tag{6-38}$$

Since the standard cell potential of the sulfate-sulfite couple is −0.093 volt, sulfite should function effectively as a depressant.

$$SO_4^{2-} + H_2O + 2e^- \rightleftharpoons SO_3^{2-} + 2OH^- \qquad E^0 = -0.93V \tag{6-39}$$

The flotation response of pyrite with ethyl xanthate as collector in the absence and presence of various additions of sodium sulfite is given in Figure 6-34. It can be noted that sulfite does function effectively in this manner as do other reducing agents.

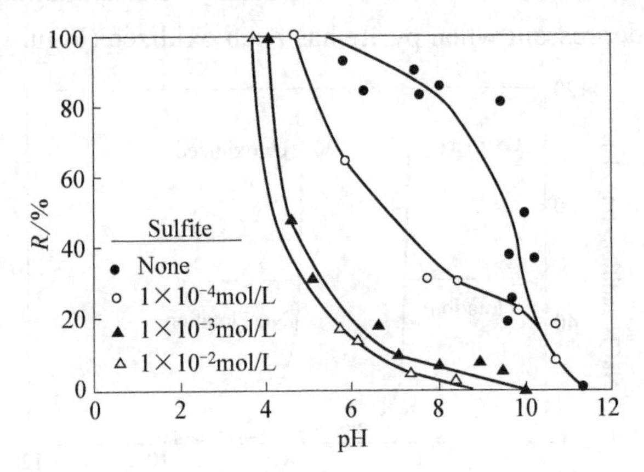

Figure 6-34 Flotation recovery of pyrite as a function of pH with 2×10^{-4} mol/L ethyl xanthate in the absence and presence of sulfite

Zeta potential determinations have also been made in this system, and chemisorption of sulfite has been noted to occur. This phenomenon also contributes to the depression of pyrite when xanthate is used as collector.

(4) Depression with sulfide. The oxidation half cell for sulfide ion is:

$$S^o + 2e^- \rightleftharpoons 2S^- \qquad E^0 = -0.48V \tag{6-40}$$

This couple is sufficiently reducing that sulfide ion should function effectively as a depressant for pyrite. In addition to this phenomenon, metal sulfides possess considerable stability so that chemisorption of sulfide ion on pyrite probably occurs similarly to sulfite ion on pyrite.

The relationship between pH and concentration of sodium sulfide necessary to prevent contact between pyrite particles and air bubbles in solutions of potassium ethyl xanthate (25 mg/L) is given in Figure 6-13. At lower values of pH, higher concentrations of sulfide ion are needed to prevent bubble contact. In alkaline medium relatively low additions are needed to accomplish depression.

6.6 Flotation Chemistry of Chalcopyrite

6.6.1 Chalcopyrite flotation

Chalcopyrite responds exceptionally well to flotation with xanthate as collector. In the presence of 1×10^{-5} mol/L ethyl xanthate, complete flotation is possible from about pH 3 to 12. See Figure 6-35. From this response, it is assumed that both electrochemical oxidation of xanthate to dixanthogen as well as chemisorption of xanthate on chalcopyrite are responsible for flotation. That is, dixanthogen is not stable above about pH 11, and if dixanthogen only were responsible for flotation, depression of chalcopyrite would occur above pH 11. See Figure 6-24.

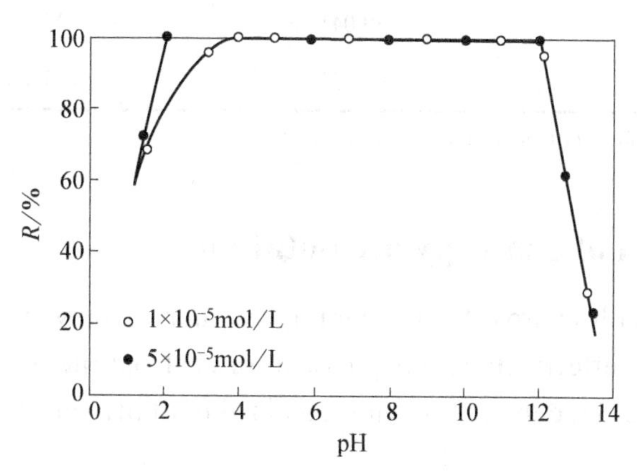

Figure 6-35 Flotation recovery of chalcopyrite as a function of pH at two different additions of ethyl xanthate

Further, in a detailed infrared spectroscopic study, Kuhn noted the presence of both dixanthogen and cuprous xanthate on chalcopyrite after contact with dixanthogen.

In an electrochemical investigation, Allison et al. found dixanthogen on chalcopyrite in contact with 6.25×10^{-4} mol/L ethyl xanthate (Table 6-1).

Abramov has also observed both dixanthogen and chemisorbed xanthate on chalcopyrite after contact with xanthate. From this investigation flotation is apparently reduced significantly under those conditions in which dixanthogen does not form on the chalcopyrite surface. However, good flotation can be achieved in the presence of chemisorbed xanthate only, but collector coatings corresponding to 4 to 5 monolayers are required.

Another sulfhydryl collector for chalcopyrite is dithiocarbamate. In an electrochemical investigation, Finkelstein and Goold observed the presence of cupric dithiocarbamate on chalcopyrite in the presence of 100×10^{-6} diethyl dithiocarbamate. See Table 6-8. The reversible potential for the oxidation of thiocarbamate to the dimer, thiuram disulfide, is relatively large, and as a result, only with pyrite is the rest potential sufficiently anodic for oxidation of collector to occur.

Table 6-8 Products of the interaction of sulfide minerals with diethyl dithiocarbamate (100×10^{-6} at pH 8)

Mineral	Rest potential/V	Product
Pyrite	0.475	Thiuram disulfide

(continued)

Mineral	Rest potential/V	Product
Covellite	0.115	Metal dithiocarbamate
Chalcopyrite	0.095	Metal dithiocarbamate
Galena	−0.035	Metal dithiocarbamate
Bornite	−0.045	Metal dithiocarbamate
Chalcocite	−0.155	Metal dithiocarbamate

Note: Reversible potential for oxidation to thiuram disulfide is 0.176.

6.6.2 Depression of chalcopyrite flotation

(1) Depression with hydroxyl. The stability of cuprous ethyl xanthate is great so that hydroxyl ion functions effectively as a depressant for chalcopyrite only at high values of pH. As is noted in Figure 6-35, depression is not obtained until pH values in excess of pH 13 are employed.

(2) Depression with cyanide. Since both electrochemical oxidation of xanthate to dixanthogen and chemisorption of xanthate can be involved in chalcopyrite flotation, it would be expected that chalcopyrite would not be as sensitive to cyanide additions as in pyrite. That this is so can be seen from the data in Figure 6-14.

(3) Depression with sulfide. In the case of sulfide ion as depressant, copper sites on the chalcopyrite surface are involved. That is, due to the stability of copper sulfide, chemisorption of sulfide ion on that site will occur.

Additionally, the oxidation potential of the solution will be lowered when sulfide ion is added. These two effects should result in chalcopyrite exhibiting greater sensitivity to sodium sulfide additions than pyrite which is what is observed (Figure 6-13).

6.7 Flotation Practice of Lead-Zinc Sulfide Ore

(1) Floatability of lead sulfide and zinc sulfide minerals

① Galena, PbS, cubic crystal, containing 86.6% Pb. It has good natural floatability.

Collector: (i) pH<9.5, the low-level xanthate is better collector of galena. When the pH value is between 7 and 8, the effect of collector is the best. It can adjust the pH value with Na_2CO_3.(ii) 9.5<pH<10.5, the advanced xanthate is better collector of galena. (iii) Galena is inhibited after pH>10.5.

The collection mechanism is chemical adsorption, and the product is lead xanthate.

Inhibitors: The inhibitors of galena are mainly sodium sulfide, dichromate, chromate, etc. Dichromate is an effective inhibitor of galena, but its inhibitory effect decreases on the activated galena by Cu^{2+}. Also, galena, which has been inhibited by dichromate, is difficult to be activated. It can only be activated after being treated with sodium chloride in hydrochloric acid or acidic medium.

Cyanide has little inhibitory effect on galena. Only some galena contaminated by iron or metamorphic galena will be inhibited by cyanide.

② Sphalerite, ZnS, containing 67.1% Zn. There are many varieties and colors of sphalerite according to its impurities. Such as brown, black (marmatite), and even colorless.

Floatability of sphalerite is good in acid medium while it needs to be activated by Cu^{2+} in alkaline medium. The most frequently used collectors are xanthate and dithiophosphate.

The most frequently used inhibitor of sphalerite is zinc sulfate, which is a relatively weak inhibitor. In addition, there are other inhibitors such as sodium sulfide, sulfite and thiosulfate, etc.

(2) Separation of lead and zinc. The separation of lead and zinc usually adopts the process of Pb preferential flotation with zinc inhibition. The inhibitors include H_2SO_3, Na_2SO_3, $Na_2S_2O_3$, $ZnSO_4$ and SO_2 gas. The main inhibiting factor is $Zn(OH)_2$, or the complex ion that may form zinc hydroxide.

(3) Separation of zinc and sulfur. There are two separation methods: the process of zinc flotation with sulfur inhibition or sulfur flotation with zinc inhibition.

Lime method is the most commonly used method of zinc flotation with sulfur inhibition. Pyrite is inhibited by adjusting the pH with lime, which is generally above 11. The advantage of this method is simple, and the reagent lime used is cheap and easy to get. The disadvantage is that the flotation equipment, especially the pipeline, is easy to scale, and the sulfur concentrate is not easy to filter.

(4) Flotation example of lead zinc ore. A high sulfur lead zinc silver polymetallic sulfide ore in China, the metal minerals in the ore are galena, sphalerite, pyrrhotite, pyrite, etc. Gangue minerals are silicate minerals such as quartz, feldspar, pyroxene, garnet, chlorite and carbonate minerals such as calcite and dolomite (Table 6-9).

Table 6-9 Results of multi-element analysis of raw ore

Element	Cu	Pb	Zn	Ag/(g/t)	Au/(g/t)	S	Fe	Mg
Content/%	0.16	1.01	4.27	39.4	<0.05	19.89	32.10	1.82
Element	P	Cd	Bi	SiO_2	As	Ca	Al_2O_3	In
Content/%	0.031	0.039	0.028	19.73	<0.005	2.72	2.78	0.0035

Based on the flotation process of Pb preferential flotation with zinc inhibition, the potential regulation flotation technology is used, that is, the regulator of lime and inhibitors of zinc sulfate and sodium sulfite are added to the mill, and the lime is used to regulate and stabilize the pulp potential to create a low oxidation potential atmosphere. At the same time, the inhibitor is added to strengthen the inhibition effect of sphalerite and pyrite minerals. The lead-zinc sulfide flotation process by potential regulation and control is shown in Figure 6-36.

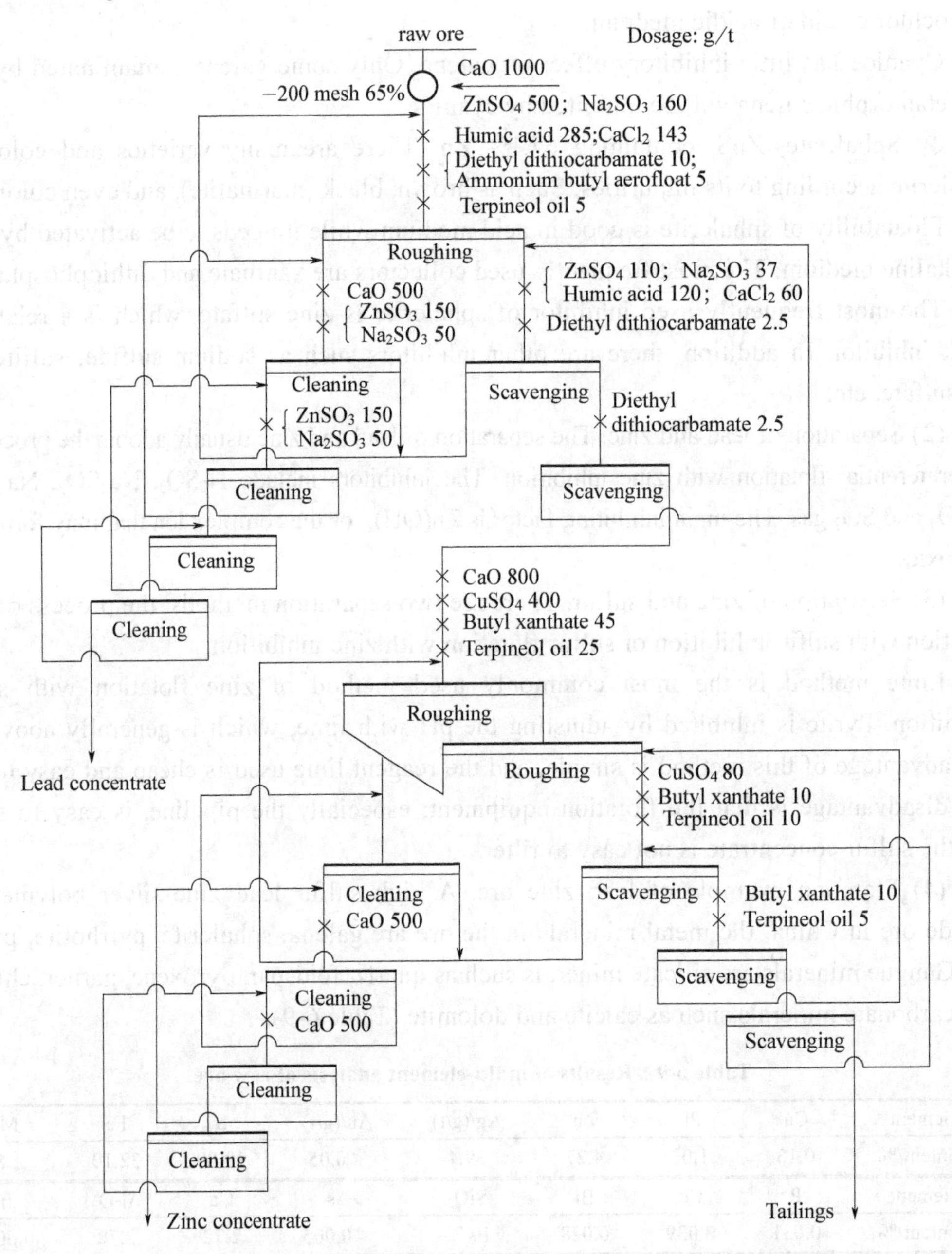

Figure 6-36 Lead-zinc sulfide floation process by potential regulation and control

The lead grade and recovery of lead concentrate were increased by 26% and 25% respectively; the zinc grade of zinc concentrate is increased by 5%. The results show that the flotation separation effect of lead and zinc is remarkable.

Chapter 7
Flotation Chemistry of Insoluble Oxide and Silicate

The number of minerals that fall into this category is large, and whether a particular mineral can be floated with a particular collector depends on the electrical properties of the mineral surface, the electrical charge of the collector, the molecular weight of the collector, the solubility of the mineral, and the stability of the metal collector salt. Depending on these phenomena, adsorption of collector may occur either by electrostatic interaction with the surface (physical adsorption) or by specific chemical interaction with surface species.

7.1 Flotation Chemistry of Physical Adsorption

7.1.1 Flotation by physical adsorption

The adsorption of many collectors occurs by electrostatic interaction with oxide and silicate surfaces. When these collectors are used, knowledge of the pzc of the minerals in question must be known. The pzc of a number of oxides and silicates are presented in Table 7-1.

Three mineral systems, the hematite, alumina and quartz systems, have received considerable attention. The dependence of hematite flotation on electrostatic phenomena when either amines or certain anionic collectors are used is shown clearly in Figure 7-1.

Such anionic collectors do not form insoluble metal-collector salts. The point-of-zero charge for hematite was measured at pH 6.7. Below the pzc the surface is positively charged; negatively charged sulfonate ions are adsorbed in this region, and complete flotation is effected. Above the pzc the surface is negatively charged, and sulfonate ions are repelled from the surface. On the other hand, aminium ions which are positively charged, are adsorbed on the negatively charged surface.

Similar phenomena can be noted when corundum (Al_2O_3) is floated with dodecylamine and dodecyl sulfate (Figure 7-2). The pzc of alumina is pH 9.

Table 7-1 pzc of several oxide and silicate minerals

Mineral	pzc (pH)	Mineral	pzc (pH)
Augite	2.7	Garnet	4.4
Bentonite	<3.0	Goethit	6.7
Beryl	3.1,3.3,4.4	Hematite	5.0,6.0,6.7
Biotite	0.4	Kaolinite	3.4
Cassiterite	4.5	Magnetite	6.5
Chromite	5.6,7.0,7.2	Pyrolusite	5.6,7.4
Chrysocolla	2.0	Quartz	1.8
Corundum	9.0,9.4	Rhodolite	2.8
Cummingtonite	5.2	Rutile	6.7
Cuprite	9.5	Tourmaline	4.0
Diopside	2.8	Zircon	5.8

Note: These are typical results. Source of oxide, trace impurities, etc., cause variations in values.

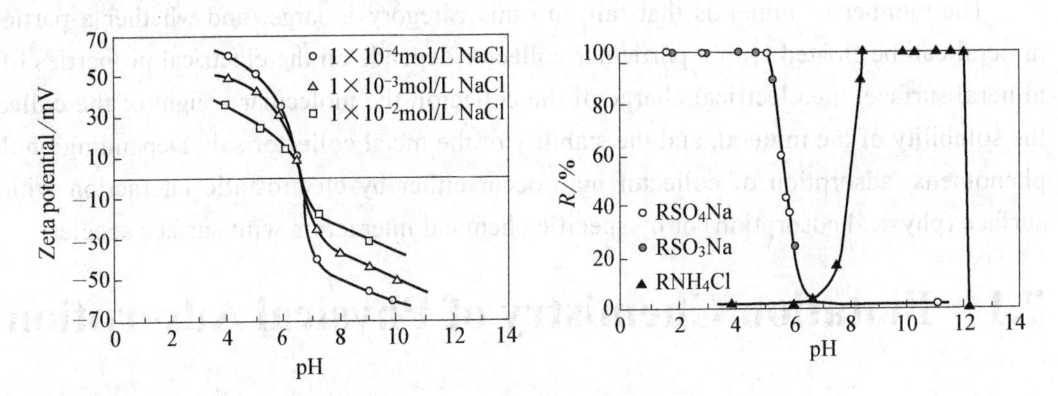

Figure 7-1 Flotation recovery of hematite as a function of pH with 1×10^{-3} mol/L additions of collector compared with surface properties of the same mineral

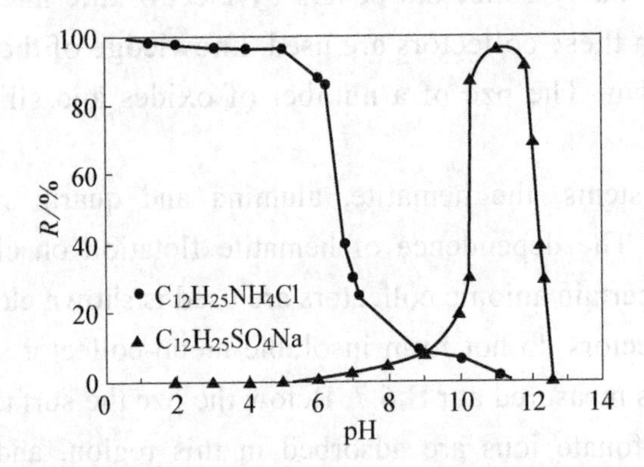

Figure 7-2 Flotation recovery of corundum as a function of pH with 4×10^{-5} mol/L additions of collector

In the case of quartz, the pzc is pH 1.8. Anionic collectors, such as sulfonates and alkyl sulfates, are not adsorbed in sufficient amount below the pzc to result in flotation, since these species must compete with anions present (from pH adjustment) in concentrations greater than 1×10^{-2} mol/L. Above the pzc aminium ions are adsorbed.

As discussed previously, extensive hydrogen bonding of water molecules occurs on oxide and silicate surfaces. As a result the presence of hemimicelles of collector appears to be necessary to render these surfaces sufficiently hydrophobic for flotation to occur. With reference to Figures 7-3 and 7-4, the concentrations at which the zeta potential of quartz changes sign from negative to positive and the concentrations at which a rapid rise in flotation recovery of quartz occurs provide evidence of this phenomenon. Some researchers have shown that when the log of collector concentration required for a rapid rise in flotation recovery is plotted as a function of the number of carbon atoms in the hydrocarbon chain (*kT* plot) a straight line with a slope which corresponds to a specific adsorption potential, φ, of -1.1 *kT* is obtained which is indicative of hemimicelle formation. This value of -1.1 *kT* corresponds to -0.62 kcal/mol CH_2 group, the free energy decrease associated with hydrocarbon chain removal from solution.

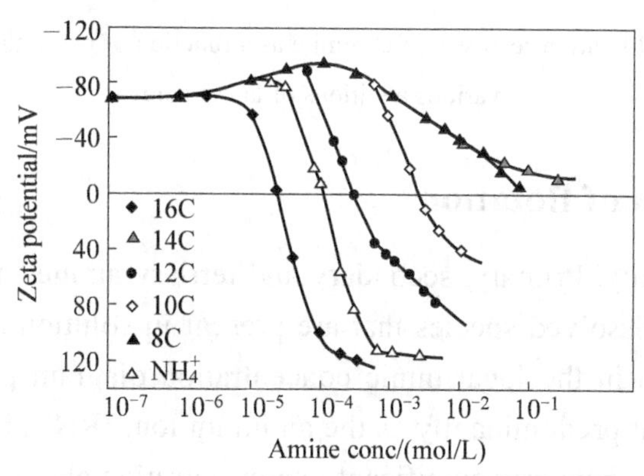

Figure 7-3 Zeta potential of quartz as a function of concentration of ammonium ion and aminium acetates with different hydrocarbon chain length at neutral pH

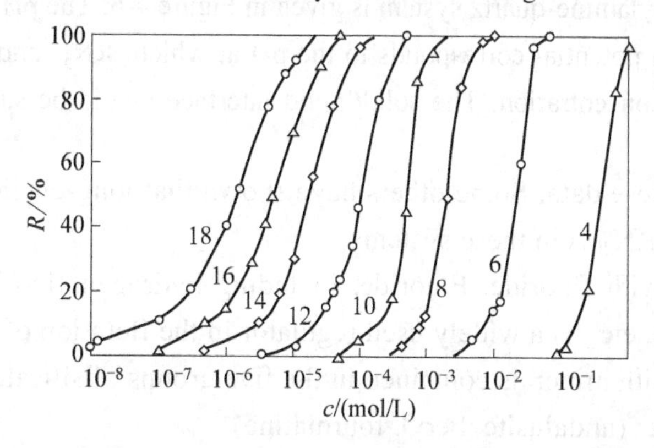

Figure 7-4 The effect of hydrocarbon chain length on the relative flotation response of quartz in the presence of aminium acetate solutions at neutral pH

Similarly to the case of dodecyl sulfonate and dodecyl sulfate flotation of hematite, the adsorption of relatively short-chain carboxylates also occurs by electrostatic interaction with oxide and silicate minerals. For example, when chromite with a pzc of pH 7.0 is floated with octanoic and decanoic acids at concentration levels of 10^{-2} molar, flotation is possible only below the pzc. When this same mineral is floated with 1×10^{-4} molar laurate, flotation also occurs only below the pzc. However, with a laurate addition of 1×10^{-3} molar, chemisorption of the collector occurs, since flotation above the pzc is possible. See Figure 7-5.

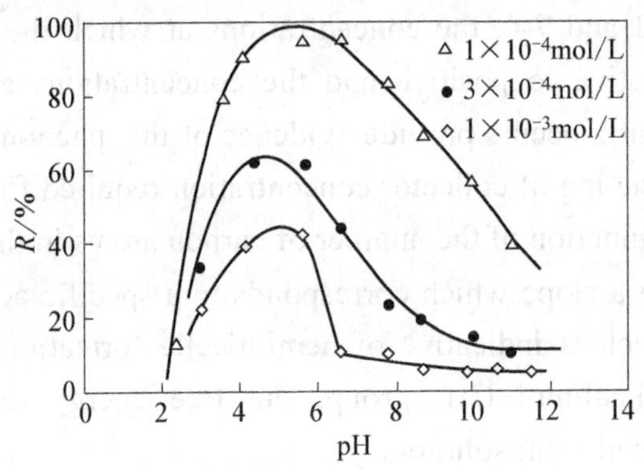

Figure 7-5 Flotation recovery of chromite as a function of pH in the presence of various additions of lauric acid

7.1.2 Modulation of flotation

(1) Effect of basicity. Primary, secondary and tertiary amines are weak bases, and the concentrations of the dissolved species that are present in solution are related to the pH of the solution. As shown in the logarithmic concentration diagram presented in Figure 3-6, dodecylamine is present predominantly as the aminium ion, RN_3^+, below about pH 10. The neutral species, $RN_{2(aq)}$, becomes significant in concentration above this acme value in pH.

The correlation between contact angle, adsorption density, zeta potential and flotation recovery for the dodecylamine-quartz system is given in Figure 7-6. The pH of maximum recovery, contact angle and zeta potential corresponds to the pH at which RN_3^+ and $RN_{2(aq)}$ are present in approximately equal concentration. The solid/liquid interface would be similar to that shown in Figure 2-7(c).

In addition to these data, Some others have shown that long-chain alcohols function in a similar capacity to $RN_{2(aq)}$ in these systems.

(2) Activation with fluoride. Fluoride, including hydrogen fluoride, sodium fluoride, sodium fluorosilicate, etc., is a widely used regulator in the flotation of silicate minerals.

Flotation data with minerals contained in the five groups of silicates has obtained.

① Orthosilicates (andalusite, beryl, tourmaline)

② Pyroxenes (augite, diopside, spodumene)

③ Amphiboles (hornblende, tremolite, actinolite)

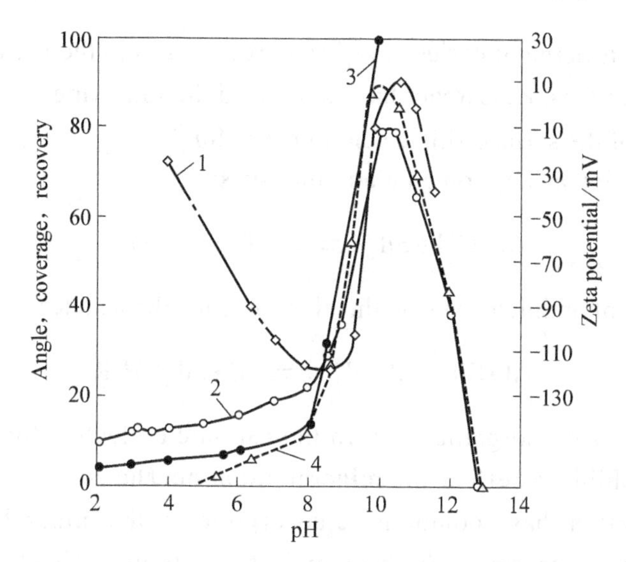

Figure 7-6 Correlation of flotation recovery, surface coverage, contact angle and zeta potential with flotation of quartz with 4×10^{-5} mol/L dodecylammonium acetate addition

1—zeta potential /mV; 2—contact angle /(°); 3—surface coverage/% (by monolayer); 4—flotation recovery /%

④ Sheet silicates (muscovite, biotite, chlorite)

⑤ Framework silicates (quartz, feldspar, nepheline)

It was found that the orthosilicates are sensitive to fluoride addition, whereas the pyroxenes and amphiboles are scarcely affected by fluoride. Sheet silicates are activated by fluoride, while framework silicates are activated to a lesser extent with the exception of quartz.

Quartz and feldspar have been floated with dodecylamine in the absence and presence of sodium fluoride. These results are presented in Figure 7-7. As shown, maximum contact angle on microcline and, hence, maximum flotation are obtained at approximately pH 2 in the presence of 10^{-2} molar fluoride.

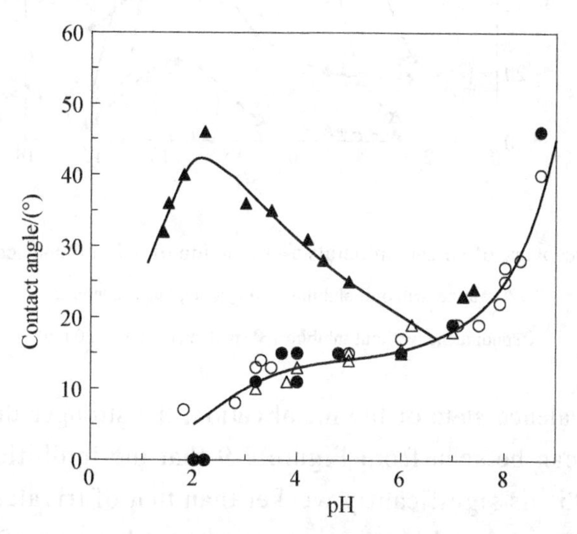

Figure 7-7 Contact angle on microcline and quartz as a function of pH with 4×10^{-5} mol/L dodecylamine in the absence and presence of fluoride

▲microcline, 10^{-2} mol/L NaF; △ microcline, no F^-; ●quartz, 10^{-2} mol/L NaF; ○quartz, no F^-

A number of conflicting theories have been presented as to the possible role of fluoride ion under these conditions. Researcher have suggested the following:

① HF attack of the surface silicic acid to form SiF_6^{2-} ,

② Fluosilicate ion adsorption on aluminum sites:

$$Al \cdot OH + SiF_6^{2-} \rightleftharpoons Al \cdot SiF_6^{-} + OH^{-} \tag{7-1}$$

③ Adsorption of aminium ions on the alumino-fluosilicate sites:

$$Al \cdot SiF_6^{-} + RNH_3^{+} \rightleftharpoons Al \cdot SiF_6NH_3R \tag{7-2}$$

(3) Depression with inorganic ions. In the cationic collector flotation system, metal cations have an inhibitory effect on mineral flotation.The metal cations and cationic collectors in the system have competitive adsorption on the mineral surface. When the concentration of metal cations is high enough, the cationic collectors can be desorbed from the mineral surface. At the same time, the adsorption of multivalent metal ions on the minerals increases the positive charge on the mineral surface, thus affecting the adsorption of cationic collectors on the mineral surface, thereby inhibiting the mineral flotation. Figure 7-8 shows the inhibitory effect of Fe^{3+} on silicate minerals in the dodecylamine system.

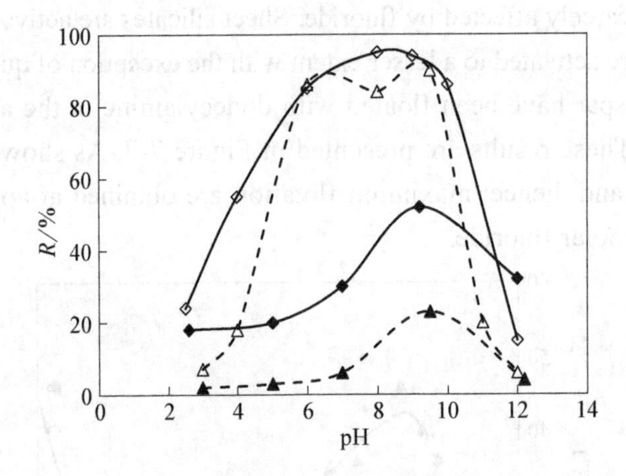

Figure 7-8 Flotation recovery of silicate minerals as a function of pH in the absence and presence of Fe^{3+}

△kyanite, without inhibitor; ▲kyanite, $FeCl_3$ 20mg/L;

◇spodumene, without inhibitor; ◆spodumene, $FeCl_3$ 20mg/L

The higher the valence state of the metal cation, the stronger the inhibition ability of mineral flotation. It can be seen from Figure 7-9 that the inhibition ability of divalent metal cations Ca^{2+}\Pb^{2+} is significantly weaker than that of trivalent metal ions Fe^{3+}. In addition, medium pH and metal ion concentration also have an effect on the inhibition of metal ions.

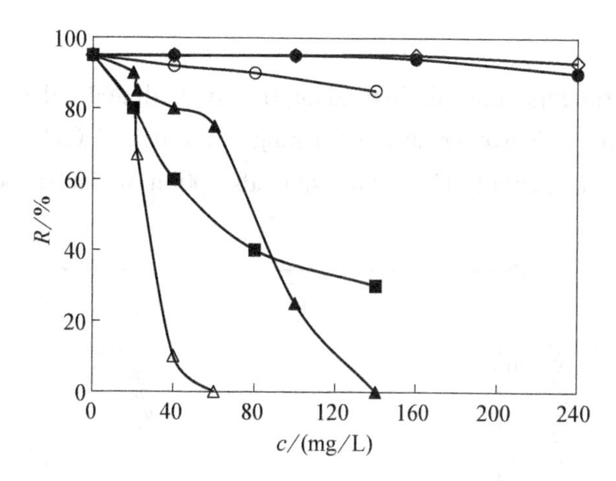

Figure 7-9 Flotation recovery of aegirine as a function of metal ion concentration with 80 mg/L dodecylamine

◇$CaCl_2$ (pH=8-9); ●$MgCl_2$ (pH=9); △$FeCl_3$ (pH=6); ▲$FeCl_3$ (pH=6, 2# oil 960 mg/L); ■$Pb(NO_3)_2$ (pH=6.8); ○$Pb(NO_3)_2$ (pH=6.8, 2# oil 720 mg/L)

7.2 Flotation Chemistry of Chemisorption

7.2.1 Flotation by chemisorption

Chemisorption of high molecular weight collectors on oxides and silicates appears to involve the hydrolysis of cations comprising these minerals. The hydroxy complexes thus formed are very surface active; they adsorb strongly on solid surfaces and reverse the sign of the zeta potential if their concentration is sufficiently high. In fact, hydroxy complexes even adsorb on positively charged surfaces.

Comparison of the pH range in which the zeta potential of quartz is positive in the presence of a metal ion with the distribution diagram for the cation illustrates these phenomena well. For example, in the case of a 1×10^{-5}mol/L lead chloride solution, $PbOH^+$ is predominant from pH 6.2 to 10.8. This is exactly the pH range in which the zeta potential of quartz is positive in the presence of 1×10^{-4}mol/L lead chloride, and this phenomenon is probably due to the adsorption of $PbOH^+$ on the surface of quartz. Compare Figure 6-12 and Figure 7-10.

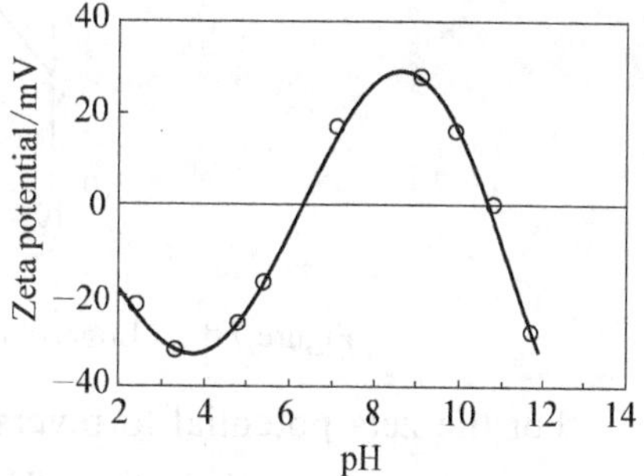

Figure 7-10 Zeta potential of quartz as a function of pH in the presence of 1×10^{-4} mol/L $PbCl_2$

Similar observations have been made in the aluminum-quartz system, the nickel-quartz system, the cupric-chrysocolla system, the manganous-rhodonite system,and the ferrous-chromite system.

Adsorption experiments show limited adsorption of hydrated calcium ion on quartz from approximately pH 4 to 11. However, in the pH range in which $CaOH^+$ is present in significant concentration namely above about pH 11, extensive adsorption of $CaOH^+$ occurs. See Figure 7-11 and Figure 7-12.

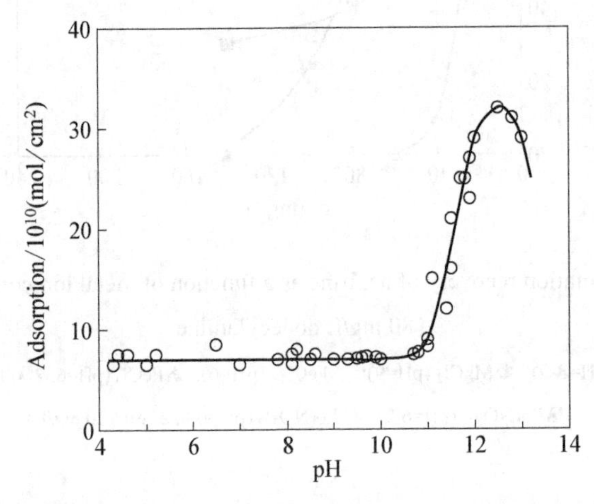

Figure 7-11 Adsorption of calcium species on quartz as a function of pH from solutions 100×10^{-6} in Ca^{2+}

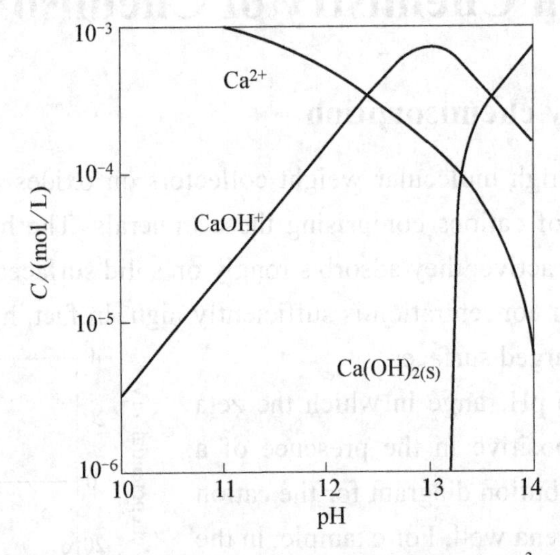

Figure 7-12 Logarithmic concentration diagram for 1×10^{-3} mol/L Ca^{2+}

For the zeta potential to reverse sign, more charge must be left in the stern layer than was present prior to adsorption. Three mechanisms of adsorption have been suggested for polyvalent cations.

These are: (1) water formation by combination of the hydroxy ion of the hydroxy complex and adsorbed hydrogen ion, (2) hydrogen bonding of the hydroxy complex with the surface, and (3) formation and adsorption of the metal hydroxide on the solid surface.

With respect to hydrogen bonding of a divalent cation hydroxy complex, using quartz as the subject mineral, hydrogen bonding could occur between adsorbed hydrogen ion and the hydroxy complex as follows:

```
\O    O·H···O—M+
  \  /      |
   Si       H
  /  \
 /O   O·H
```

An alternative mechanism is the adsorption of the hydroxy complex by the formation and splitting out of water:

```
\O    O  H···O—M+            \O    O·M+
  \  /        |                \  /
   Si         H      ——→        Si          + H2O
  /  \                         /  \
 /O   O·H                     /O   O·H
```

A third mechanism has been suggested involving nucleation and growth of a hydroxide precipitate at the surface. Following precipitate formation, the surface is postulated to acquire the electro kinetic features of the metal hydroxide as shown in Figure 7-13. Any of these mechanisms would result in more charge in the stern layer following adsorption. At the present time, it is not possible to state which of these mechanisms may be operable.

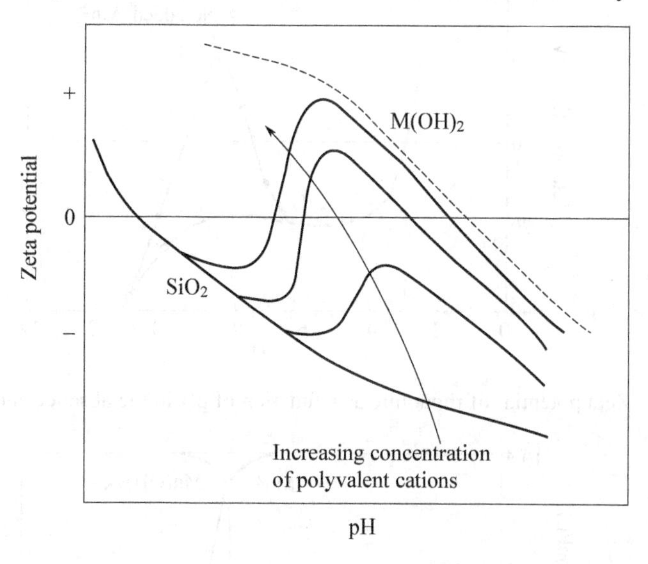

Figure 7-13 Zeta potential of quartz as a function of pH and divalent metal ion concentration

As mentioned previously, chemisorption of high molecular weight collectors on oxides and silicates occurs under conditions where metal ions comprising the mineral have hydrolyzed to hydroxy complexes. When pyrolusite is floated with oleate as collector, two maxima in flotation response are present (Figure 7-14). The pzc of this mineral is pH 7.4. From electrophoretic and activation work in other systems, the flotation response observed at pH 8.5 is intimately involved with the species $MnOH^{+}$. This may be seen by referring to Figure 7-15 which shows the zeta potential of rhodonite (monogamous silicate) in the absence and presence of manganese sulfate and also to Figure 7-16—the distribution diagram of manganese ion. As can be noted, zeta reversal and flotation occur in the pH range in which $MnOH^+$ is present maximally. Further, this is the same pH range in which rhodonite responds to flotation with octyl hydroxamate and activation of quartz results in the presence of manganous salts (Figure 7-36).

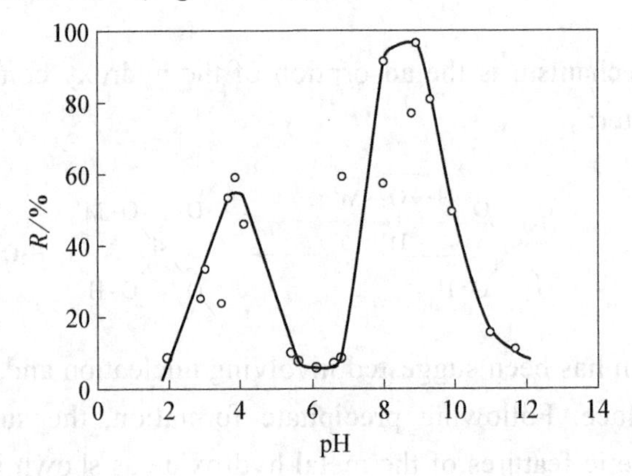

Figure 7-14 Flotation recovery of pyrolusite as a function of pH with 1×10^{-4} mol/L oleate

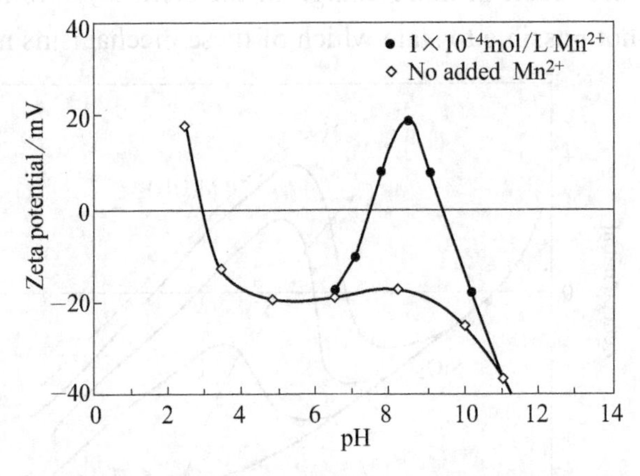

Figure 7-15 Zeta potential of rhodonite as a function of pH in the absence and presence of Mn^{2+}

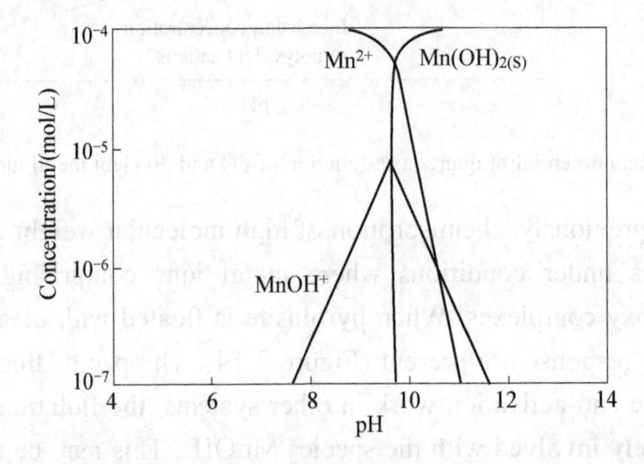

Figure 7-16 Logarithmic concentration diagram for 1×10^{-4} mol/L Mn^{2+}

Pyrolusite is manganic oxide, and formation of the hydroxy complex of manganous ion could occur only after manganic reduction by reaction with water. In practice dissolution of pyrolusite has been enhanced by the addition of sulfur dioxide to ore systems.

Flotation observed at pH 4 can probably be attributed to physical adsorption of an oleate species. Unfortunately, pyrolusite is opaque to light in the infrared region, and identification of species present on the surface cannot be made.

The role that hydroxy complexes assume in chemisorptions in anionic flotation systems is also clearly illustrated in the chromite and augite systems. The theoretical composition of chromite is $FeO \cdot Cr_2O_3$, but as found in nature, Mg^{2+} is frequently substituted for Fe^{2+}, while Fe^{2+} and Al^{3+} are substituted for Cr^{3+}.

As would be expected, composition exhibits a direct effect on the electrical properties of this mineral's surface. In fact, pzc ranging from pH 4.4 to 9.6 have been measured for chromites of various compositions. The composition of the chromite whose flotation response is given in Figures 7-5 and 7-17 is 41.72% Cr^{3+}, 7.05% Fe^{2+}, 3.73% Fe^{3+}, 8.94% Mg^{2+}, 4.21% Al^{3+}, 0.11% Ca^{2+}, 0.31% Si^{4+}, and 0.07%Ti^{4+}.

The response in the vicinity of pH 11 is attributed to $MgOH^+$ and that in the vicinity of pH 8 is attributed, in pact, to $FeOH^+$. These are the pH values at which zeta reversal occurs and those at which hydroxy complexes are present in significant concentration. See Figures 7-18 and 7-19.

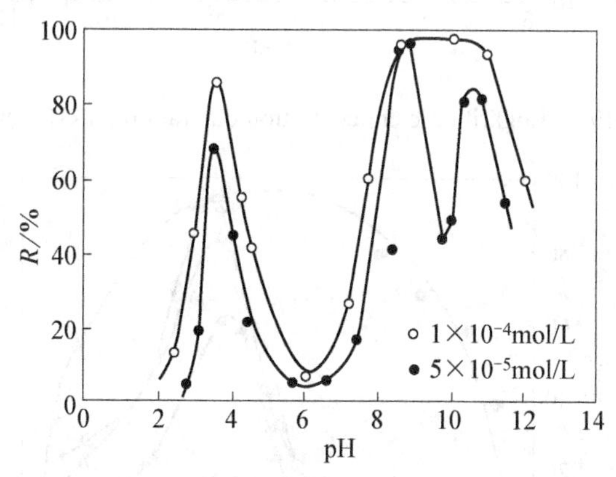

Figure 7-17 Flotation recovery of chromite as a function of pH and oleate concentration

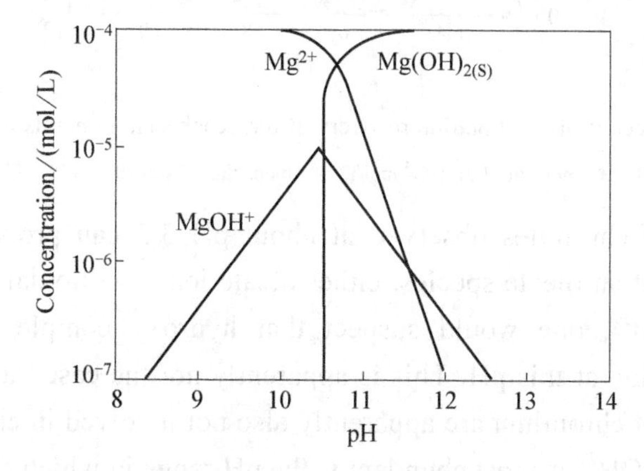

Figure 7-18 Logarithmic concentration diagram for 1×10^{-4}mol/L Mg^{2+}

Researchers have shown that non-iron-bearing minerals are also floated at approximately pH 8 as shown in Figure 7-20. Further others have shown that hematite is floated maximally at the same pH with oleic acid. Apparently, phenomena in addition to hydro complex formation are involved in the flotation response observed at pH 8. For example, hematite flotation with oleate that collector adsorption at the air/water interface and subsequent lowering of surface tension may also involved in maximal flotation observed at this pH. The dynamic surface tension in such systems seems to be related to acid soap formation. A theoretical explanation for this surface tension lowering has been presented.

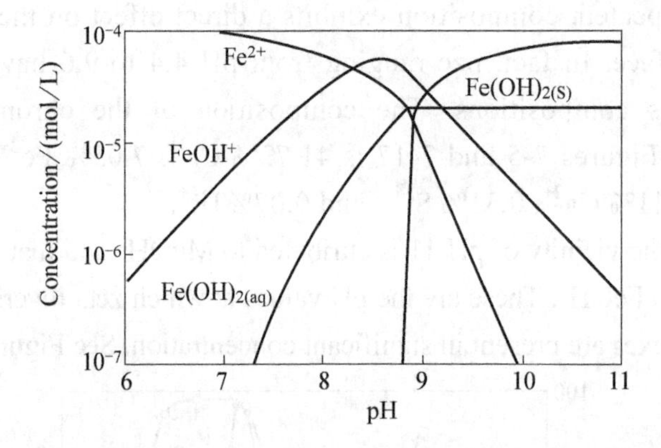

Figure 7-19 Logarithmic concentration diagram for 1×10^{-4} mol/L Fe^{2+}

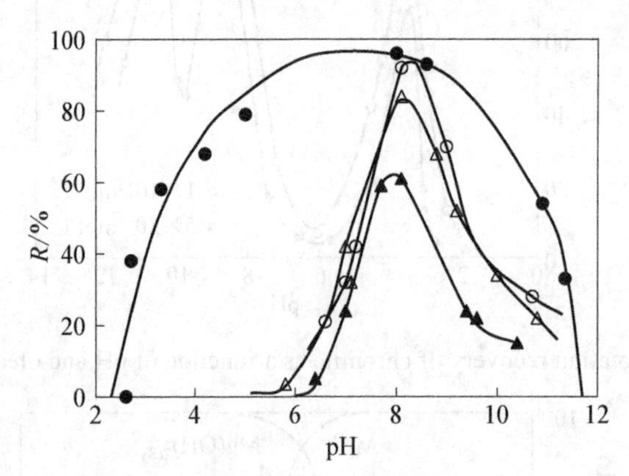

Figure 7-20 Effect of pH on flotation recovery of the beach sand minerals with sodium oleate

● zircon 8.42×10^{-5} mol/L; ▲ monazite 1.30×10^{-4} mol/L; △ monazite 2.60×10^{-4} mol/L; ○ ilmenite 3.18×10^{-4} mol/L

The flotation of chromites observed at about pH 3.5 can probably be attributed to physical adsorption of an oleate species, either oleate ion or colloidal oleic acid. Since this chromite contains Al^{3+}, one would suspect that hydroxy complexes of aluminum are responsible for flotation at this pH. This is apparently not the case, however. Similarly, the hydroxy complexes of chromium are apparently also not involved in chromite flotation. The pH range in which $CrOH^{2+}$ is most abundant is the pH range in which depression is observed. Aluminum and chromium are coordinated octahedrally with oxygen, whereas the divalent cations are coordinated tetrahedrally with oxygen. As a result the divalent cations would be expected to dissolve much more readily than the trivalent cations.

Response similar to these was also observed. With augite, the response observed at pH 3 can probably be attributed to physical adsorption of an oleate species. See Figure 7-21 and Figure 7-22.

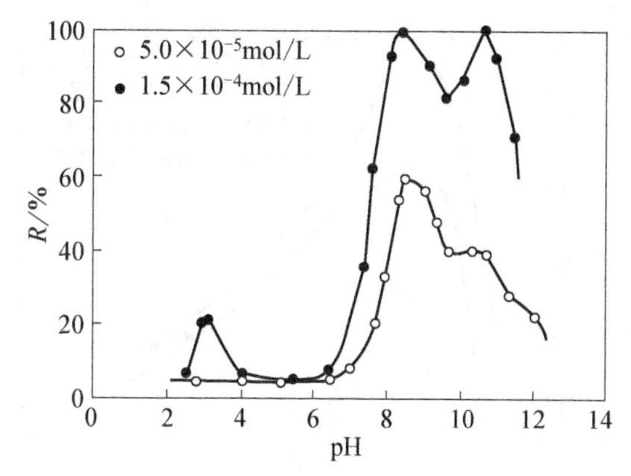

Figure 7-21 Flotation recovery of augite as a function of pH and oleate concentration

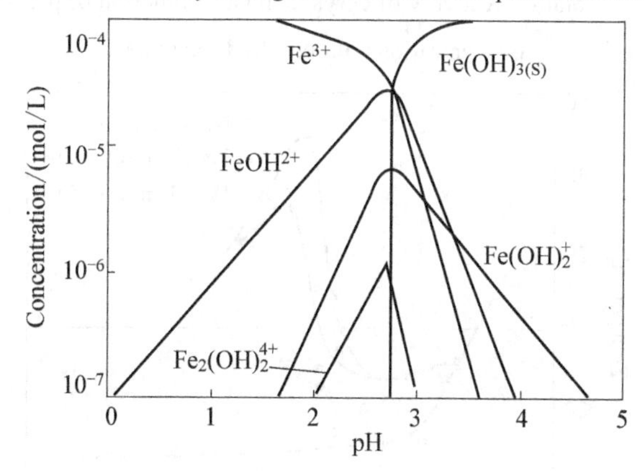

Figure 7-22 Logarithmic concentration diagram for 1×10^{-4} mol/L Fe^{3+}

Since the pzc of this mineral is pH 2.7, the response of this mineral is much more limited than that observed for chromites and pyrolusite under the same conditions. The pzc of chromites and pyrolusite are pH 7.0 and pH 7.4, respectively. The responses observed at around pH 8 may be attributed to $FeOH^+$ and that from pH 10 to pH 12 to $MgOH^+$ and $CaOH^+$. The distribution diagrams for calcium, ferrous ion and magnesium are presented in Figures 7-12, 7-18 and 7-19.

Chrysocolla responds to flotation with octyl hydroxamate in the pH range in which $CaOH^+$ is present in significant concentration. Optimum flotation occurs at approximately pH 6.3 which is also the pH at which the zeta potential becomes positive in the presence of cupric salts. See Figures 7-23 and 7-24.

As opposed to pyrolusite, chrysocolla responds well to infrared absorption analysis. Major absorption bands of cupric octyl hydroxamate occur at 925, 1095,1380,1450 and 1535 cm^{-1}. Identical absorption bands at 1380, 1450 and 1535 cm^{-1} are noted on chrysocolla after contact with 1×10^{-3} molar octyl hydroxamate at pH 6.3. In fact, the multilayers of cupric hydroxamate are visible apparent on the surface. The chrysocolla changes color from its natural blue-green to a vivid green color during the conditioning step prior to flotation. Adsorption of multilayers of cupric hydroxamate, having formed on or very near the surface, could occur on previously chemisorbed hydroxamate via hydrocarbon chain association.

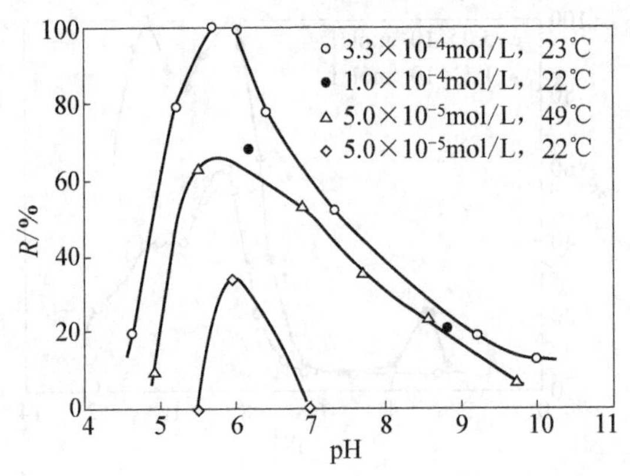

Figure 7-23 Flotation recovery of chrysocolla as a function of pH with various concentrations of octyl hydroxamate

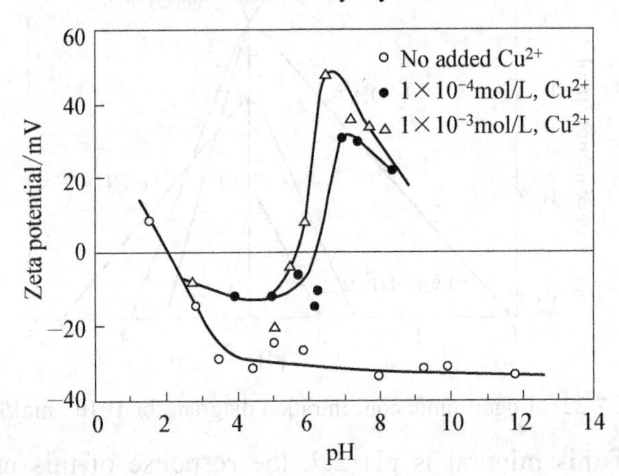

Figure 7-24 Zeta potential of chrysocolla as a function of pH in the absence and presence of Cu^{2+}

Multilayers of metal hydroxamate have been observed in other systems, notably the hematite-hydroxamate system. In the case of pyrolusite, multilayer adsorption of octyl hydroxamate was observed, but whether manganous hydroxamate was present as multilayers was not established.

The beryllium silicate minerals, beryl and phenacite, offer additional evidence of the necessity of slight mineral dissolution for collector adsorption and flotation. Beryl, $Be_3Al_2(SiO_3)_6$, is known to be very insoluble and is not concentrated by oleate flotation without activation by polyvalent cations, particularly ferric. On the other hand, phenacite, Be_3SiO_4, apparently exhibits easer solubility and can be concentrated by oleate flotation, as shown in Figure 7-25. Beryllium hydrolyzes to its hydroxy complexes in the pH range, as

flotation to occur. The chemisorption of oleate was established by infrared spectroscopy.

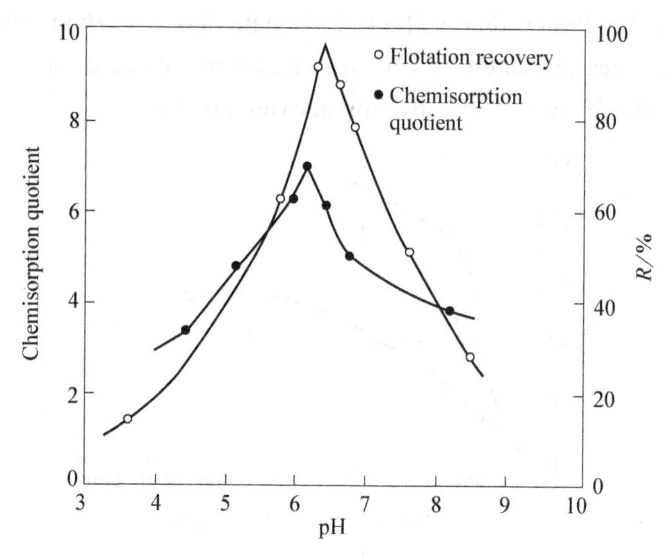

Figure 7-25 Chemisorption of oleate and flotation recovery of phenacite

7.2.2 Modulation of flotation

(1) Effect of basicity. Sulfuric acid, hydrochloric acid, sodium hydroxide, sodium carbonate, etc. are all important pH adjusters. While changing the pH value of pulp, they are often intertwined with the activation and inhibition of mineral flotation. The addition of these agents can affect the behavior of collectors and unavoidable ions in the pulp, thereby affecting the flotation of minerals. In addition, these agents have chemical cleaning or pretreatment effects on the mineral surface.

The role of sodium hydroxide in silicate mineral flotation is not only a pH regulator, but also a pretreatment reagent. After some minerals are treated with sodium hydroxide, the impurities and slime on the surface of these minerals can be reduced and eliminated, the uneven dissolution on the surface of these minerals can be generated, the silicate surface area with strong hydration can be reduced, and the metal cations can be enriched on the surface of these minerals, which is conducive to the adsorption of anion collectors on the surface of these minerals. Figure 7-26 shows the floatability of spodumene at different oleic acid dosage after sodium hydroxide pretreatment. Figure 7-26 shows that spodumene is easier to float after sodium hydroxide activation.

Sodium carbonate is an extremely important alkaline regulator in the flotation of silicate minerals with fatty acid collectors. Fatty acid collectors are most effective within the stable pH range created by sodium carbonate. Sodium carbonate can also eliminate the harmful effects of Ca^{2+}, Mg^{2+} and improve the selectivity of the flotation process. Sodium carbonate is also a good dispersant for slime, which can prevent the agglomeration of fine slime in the pulp and also improve the selectivity of flotation process. Figure 7-27 shows the effect of sodium carbonate on the floatability of spodumene and beryl activated by Ca^{2+}. It can be seen from the figure that sodium carbonate has a certain inhibitory effect on spodumene that has been activated by Ca^{2+}, but under the tested dosage, sodium carbonate has not yet shown inhibition on beryl strongly activated by Ca^{2+}.

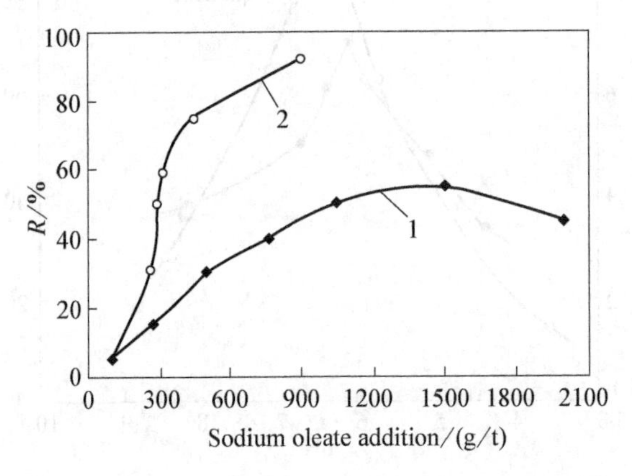

Figure 7-26 Relationship between spodumene flotation recovery rate and sodium oleate dosage after sodium hydroxide pretreatment

1—not treated; 2—pretreated with sodium hydroxide

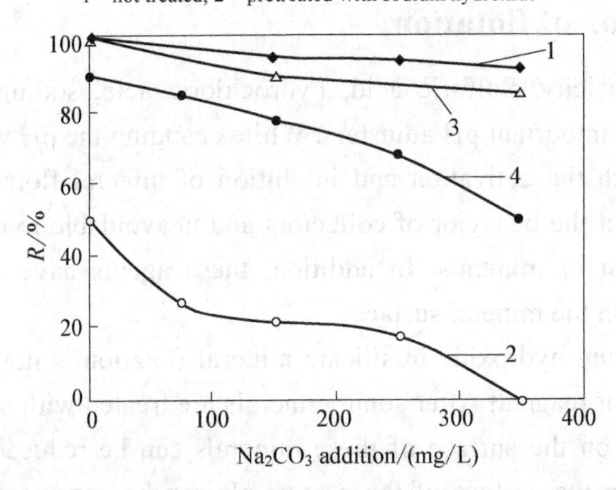

Figure 7-27 Flotation recovery of beryl and spodumene as a function of Na_2CO_3 addition at pH 11.5

Addition: 1—beryl, $CaCl_2$ 50mg/L, sodium oleate 60 mg/L; 2—spodumene, $CaCl_2$ 50mg/L, sodium oleate 60 mg/L; 3—spodumene, $CaCl_2$ 15mg/L, sodium oleate 20 mg/L; 4—beryl, $CaCl_2$ 10mg/L, sodium oleate 10 mg/L

Sodium carbonate, the same as sodium hydroxide, is also a good pretreatment agent. For example, during the flotation separation of feldspar and quartz, a certain amount of sodium carbonate is added to the pulp for stirring and scrubbing, which promotes the selective corrosion of quartz and feldspar minerals and increases the difference between the physical and chemical properties of the two minerals. The corrosion reaction of sodium carbonate in the surface of quartz is:

$$Na_2CO_3 + nSiO_2 \longrightarrow Na_2O \cdot nSiO_2 + CO_2 \uparrow \tag{7-3}$$

$Na_2O \cdot nSiO_2$ dissolved from the quartz particles can react with sulfuric acid to produce polysilicic acid. Polysilicic acid plays a very important role in the selective inhibitor of quartz and the adsorption of collector on feldspar.

(2) Activation with fluoride. In the sodium oleate flotation system, HF can inhibit most silicate minerals, but it can activate the ring structure mineral beryl, framework structure mineral quartz and double chain structure mineral ordinary amphibole under certain medium conditions (especially neutral and weak alkaline conditions). Among them, the activation of beryl is the strongest, the activation of quartz is the second, and the activation of ordinary amphibole is weak. Moreover, HF has little effect on the island structure mineral iron aluminum garnet under acidic conditions, and has weak inhibitory effect on kyanite.

NaF can slightly activate some silicate minerals (frame, ring and layered structure minerals) under alkaline conditions, and inhibit silicate minerals under other medium conditions. Compared with the effect of HF on the floatability of silicate minerals, NaF has a stronger inhibitory effect on the island structure mineral kyanite, and a weaker activation effect on the framework structure mineral quartz and the ring structure mineral beryl than HF. NaF has a good inhibitory effect on ordinary amphibole, while HF has only a slight activation effect.

The effect of Na_2SiF_6 on the floatability of silicate minerals is similar to that of HF. Na_2SiF_6 has an inhibitory effect on most minerals, but has a strong activation effect on the framework mineral quartz and the ring structure mineral beryl, especially in weakly acidic conditions, the minerals can be completely recovered. Similar to HF, Na_2SiF_6 has little effect on the island structure mineral iron aluminum garnet under acidic and weakly alkaline conditions.

The difference is that the inhibitory effect of Na_2SiF_6 on kyanite and amphibole is stronger than that of HF, and the activation effect of Na_2SiF_6 on quartz and beryl is also stronger than that of HF.

(3) Depression with inorganic ions. It can be seen from the content of the previous section that metal salt ions are often used as activators for this kind of minerals flotation by chemical adsorption, while inorganic anions can change the ion and molecular composition in the pulp. Anions can be electrostatically adsorbed on the surface of silicate minerals; ions such as silicate and sulfate can also bond with metal cations on the mineral surface by chemical bonding force, resulting in chemical adsorption or surface reaction.

Thus, hydrophilic compound films, ion adsorption films or hydrophilic colloidal particles are formed on the surface of minerals, which have an inhibitory effect on minerals. It can also dissolve the hydrophobic covering film formed by the collector on the mineral surface, so that the collector can be desorbed from the mineral surface, or form competitive adsorption with the collector on the mineral surface. When the adsorption strength and concentration of inorganic ions are sufficient, the collector can be desorbed, and a hydrophilic film of inorganic ions can be formed on the mineral surface. As discussed in the previous section, some anions can increase or decrease the active sites of collector adsorption on the mineral surface.

Sodium silicate is an effective inhibitor of silicate gangue minerals and has a good dispersing effect on slime. Water glass is a strong base and weak acid salt, which is prone to strong hydrolysis reaction in water. The ϕ-pH diagram of the aqueous solution of its component Na_2SiO_3 is shown in Figure 7-28. It has a good deactivation effect on silicate minerals activated by metal cations. The effect of Na_2SiO_3 on the floatability of silicate minerals is shown in Figure 7-29.

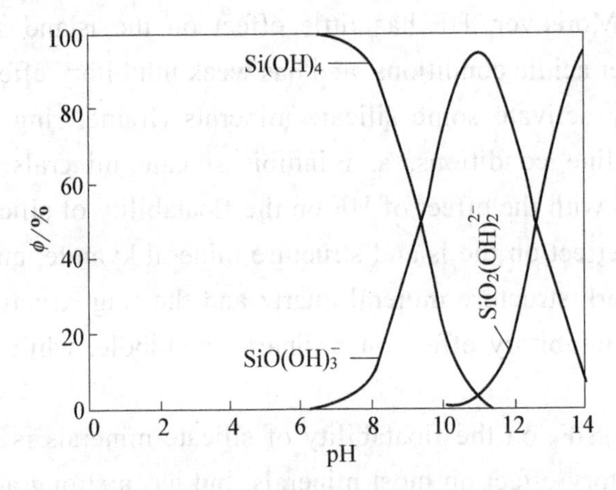

Figure 7-28 ϕ-pH diagram of Na_2SiO_3 aqueous solution

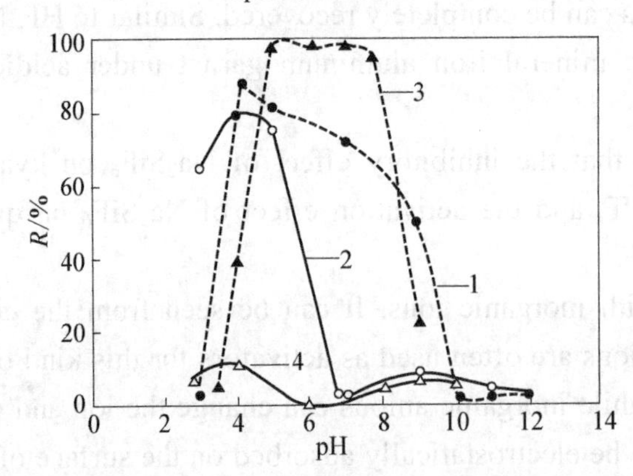

Figure 7-29 Flotation recovery of silicate minerals as a function of pH with 160 mg/L sodium oleate in the absence and presence of Na_2SiO_3

Dosage of sodium oleate: 160 mg/L; 1—kyanite, no addition; 2—kyanite, Na_2SiO_3 80mg/L ; 3—lepidolite, $AlCl_3$ 20mg/L; 4—lepidolite, $AlCl_3$ 20 mg/L+Na_2SiO_3 80mg/L

Figure 7-29 shows that under strong acid conditions, Na_2SiO_3 can activate silicate mineral kyanite, and under other medium conditions, Na_2SiO_3 has a good inhibitory effect on minerals. Especially under the alkaline condition of pH>7, the inhibitory effect of Na_2SiO_3 is very strong.

It is generally believed that under weakly acidic and alkaline conditions, Na_2SiO_3 aggregates through oxygen linkage to form negative silicic acid colloids and $SiO(OH)_3^-$, which adsorb on the surface of minerals, resulting in the inhibition of Na_2SiO_3 on minerals. When pH>7, Na_2SiO_3 has a strong inhibitory effect on minerals because the content of $SiO(OH)_3^-$ in the solution and the adsorption amount on the mineral surface are greatly increased under this condition. Under strong acid conditions, Na_2SiO_3 mainly exists in the form of silicic acid and forms positively charged silicic acid colloidal particles through hydroxyl coupling reaction polymerization. The adsorption of polymers on the surface of silicate minerals makes the minerals positively charged, resulting in the activation of Na_2SiO_3 on the flotation of silicate minerals.

Na_2SiO_3 can deactivate silicate minerals after Al^{3+} activation. Figure 7-29 also shows that Na_2SiO_3 has a good deactivation effect on lepidolite activated by Al^{3+} under the condition of pH>4. This is because under the condition of pH>4, the negatively charged silicic acid colloidal particles and $SiO(OH)_3^-$ can be adsorbed on the metal ions such as Al^{3+} or Ca^{2+} that have been adsorbed on the mineral surface, or adsorbed these metal cations to change the mineral surface electrical properties, and improve the hydrophilicity of these minerals.

The effect of Na_2SiO_3 dosage on the flotation recovery of silicate minerals when naphthenic acid dithiophosphate is used as collector is shown in Figure 7-30. It can be seen from Figure 7-30 that Na_2SiO_3 inhibits zircon, feldspar and quartz when naphthenate black drug is used as collector, of which quartz is the most easily inhibited, and Na_2SiO_3 has selective inhibitory effect on zircon, feldspar and quartz. Therefore, when the dosage of Na_2SiO_3 is 700 g/t, zircon can be separated from quartz and feldspar by flotation using naphthenic acid black drug.

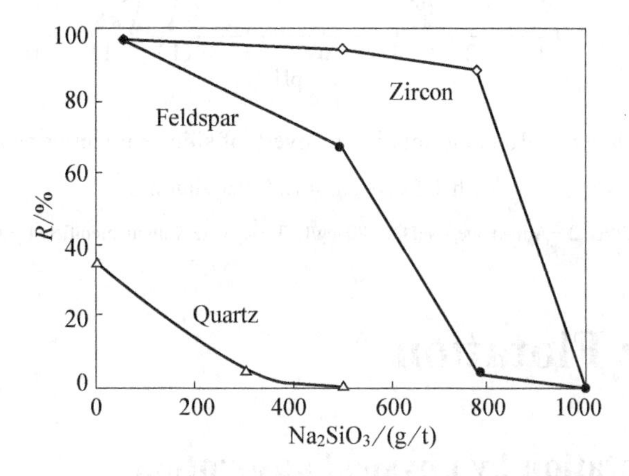

Figure 7-30 Flotation recovery of feldspar, zircon and quartz as a function of Na_2SiO_3 dosage with 1.5×10^{-4} mol/L naphthenic acid dithiophosphate

Sodium hexametaphosphate can be ionized into various anions in aqueous solution. It has strong activity and can react with metal ions in solution to form stable complexes. Its reaction with calcium ion in solution:

$$(NaPO_3)_6 \rightleftharpoons Na_4P_6O_{18}^{2-} + 2Na^+ \tag{7-4}$$

$$Na_4P_6O_{18}^{2-} + Ca^{2+} \rightleftharpoons CaNa_4P_6O_{18} \tag{7-5}$$

Sodium metaphosphate is hydrolyzed to form metaphosphoric acid, which is hydrolyzed to orthophosphoric acid. The dissociation of orthophosphoric acid is shown in Figure 3-8(a), and each ion component produced can have different inhibitory effects on flotation.

Sodium hexametaphosphate reacts with Ca^{2+}, Mg^{2+}, Fe^{2+}, Ni^{2+} and other ions to form hydrophilic and stable complexes, which enable it to inhibit minerals containing these metal cations in the crystal lattice. The adsorption of sodium hexametaphosphate on the mineral surface can also increase the negative electricity of the surface, which makes the mineral and other negatively charged minerals separate from each other under the action of electrolytic repulsion. It can make the negatively charged slime in suspension and eliminate their adverse effects on the flotation of other minerals. figure 7-31 is the effect of sodium hexametaphosphate on the flotation of silicate minerals. It can be seen from the figure that sodium hexametaphosphate has a good inhibitory effect on almandine and kyanite.

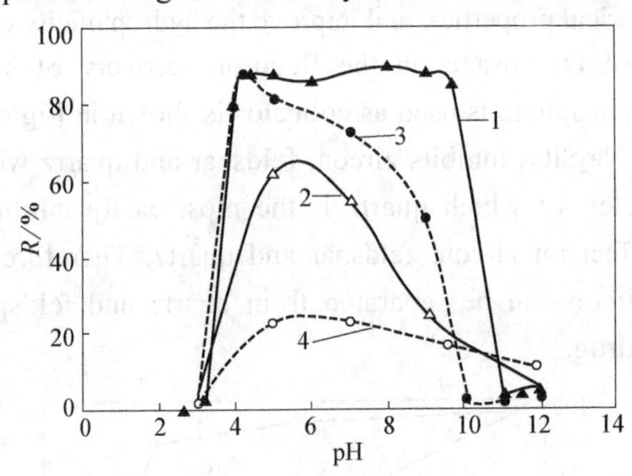

Figure 7-31 $(NaPO_3)_6$ effect on flotation recovery of silicate minerals as a function of pH with 1.5×10^{-4} mol/L dodecylamine

1—Almandine, without modifier; 2—Almandine, $(NaPO_3)_6$ 80 mg/L; 3—Kyanite, without modifier; 4—Kyanite, $(NaPO_3)_6$ 80 mg/L

7.3 Quartz Flotation

7.3.1 Quartz flotation by physical adsorption

(1) Effect of pH regulator. Quartz is usually gangue mineral. The PZC of quartz and silicate minerals is low and negatively charged in a wide range of pH values, while the PZC of useful minerals, such as hematite and apatite, is often greater than that of quartz and silicate.

Therefore, by pH adjustment, quartz and silicate minerals are negatively charged, while other minerals are positively charged. Under these conditions, amine collectors can be used for quartz flotation. The application of amine collector in the flotation of quartz and silicate minerals has been widely studied. Figure7-32 shows the effect of dodecylamine dosage and pH value on the

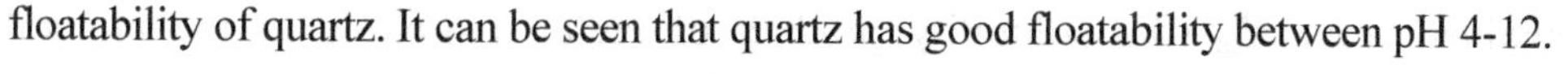
floatability of quartz. It can be seen that quartz has good floatability between pH 4-12.

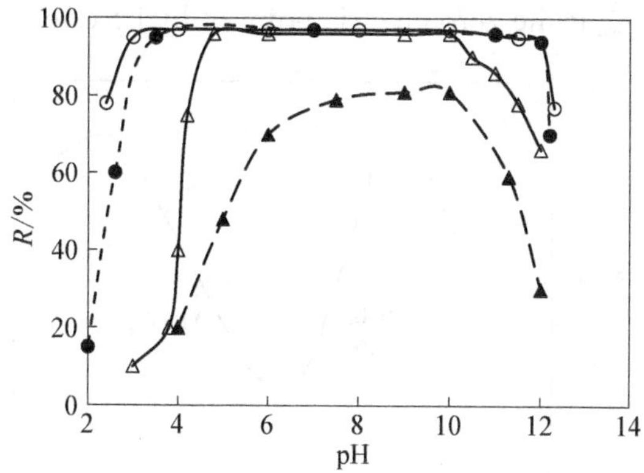

Figure 7-32 Flotation recovery of quartz as a function of pH with different dosages of dodecylamine

○ 3.0×10^{-4} mol/L; ● 2.0×10^{-4} mol/L; △ 5.0×10^{-5} mol/L; ▲ 2.5×10^{-5} mol/L

(2) Depression of quartz. Metal cations are inhibitors of quartz flotation with amine collectors. The inhibition ability of metal cations on quartz increases with the increase of its ionic valence. This is because the adsorption of multivalent metal cations on the mineral surface increases the positive charge on the mineral surface, which is consistent with the activation ability of metal cations on quartz when anionic collectors are used for flotation.

The depressing effect that inorganic cations exhibit in cationic flotation systems has been studied in detail. The flotation of quartz with 1×10^{-4} mol/L dodecylamine as a function of potassium nitrate concentration and pH is presented in Figure 7-33. As the pH is reduced, the surface charge is reduced, and the concentration of K^+ needed to depress the system is reduced. Under these conditions K^+ is able to compete effectively with RNH_3^+ for the surface.

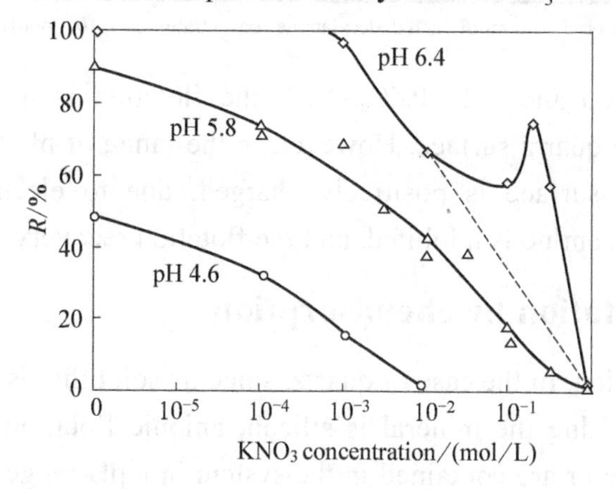

Figure 7-33 Flotation recovery of quartz as a function of KNO_3 concentration and pH with 1.5×10^{-4} mol/L dodecylamine

Dodecylamine can float quartz well between pH=2-12, but in the presence of multivalent metal cations, the floatability of quartz is inhibited. This inhibition law can be seen in Figure 7-34. CR_2 and CR_3 in the figure represent the pH value when the mineral zeta potential changes sign. pH_s is the pH value when the surface hydroxide is precipitated, which is obtained from K_{sp} (the solubility product of metal hydroxide in the mineral interface region), and PZC_e is the zero electric point of hydroxide solids.

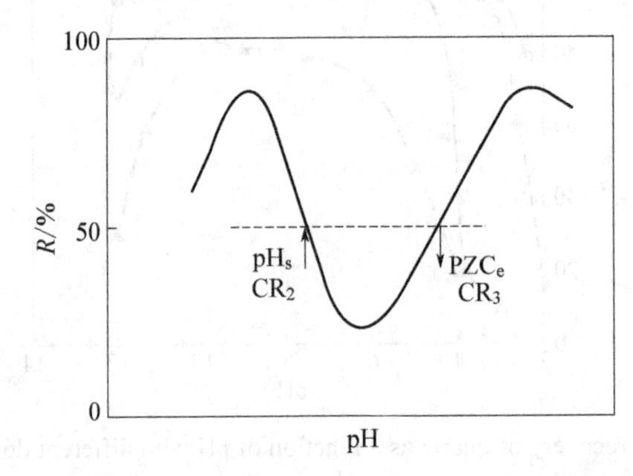

Figure 7-34 Flotation recovery of quartz as a function of pH with amine collectors

A trough appears on the flotation recovery curve, which is caused by various metal ions in the flotation of quartz with amine collectors. The pH range of the trough of the recovery curve is listed in Table 7-2.

Table 7-2 pH range of quartz flotation with dodecylamine inhibited by metal ions

Metal ion	Cu^{2+}	Zn^{2+}	Pb^{2+}	Mn^{2+}	Fe^{2+}	Co^{2+}	Ni^{2+}	Mg^{2+}	Fe^{3+}	Al^{3+}
pH_s	5.2	6.6	7.4	8.6	7.3	7.4	7.3	9.3	1.4	3.1
PZC_e	9.8	9.5	10	12.1	11.2	11.3	10.6	12.1	7.8	9.1
pH range	5.2-9.8	6.8-9.5	7.6-10.0	8.6-12	7.6-11.2	7.4-11.3	7.4-10.6	9.4-12	2.0-7.8	3.0-10.0

Note: metal ion concentration is 5×10^{-4} mol/L, and dodecylamine concentration is 5×10^{-5} mol/L.

At $pH>PZC_e(CR_3)$ and $pH<PZC_e(CR_3)$, the flotation is not inhibited due to the negative charge on the quartz surface. However, in the range of $pH_s\leqslant pH\leqslant PZC_e$ or $CR_2\leqslant pH\leqslant CR_3$, the quartz surface is positively charged, due to electrostatic repulsion, the flotation of quartz with amine is inhibited, and the flotation recovery curve shows a trough.

7.3.2 Quartz flotation by chemisorption

(1) Quartz activation. In the case of quartz, since its solubility is very limited, and since the only cation comprising the mineral is silicon, anionic flotation is obtained only after metal ions are added to or are contained in the system in a pH range in which hydrolysis to first hydroxy complexes occurs. Quartz activation can be achieved with most anionic collectors, including xanthate, providing certain conditions are satisfied.

In the case of sulfonate flotation, edges minimum values of pH at which flotation is obtained in the presence of 1×10^{-4} mol/L sulfonate and 1×10^{-4} mol/L of various metal ions are presented in Figure 7-35. Comparison of these data with the distribution diagrams of these metal ions (Figures 6-12, 7-12, 7-16, 7-19 and 7-22) should be made. The range of pH in which flotation is obtained in the presence of some of these activators is shown in Figure 7-36. The absence of flotation above the upper flotation edge and below the lower flotation edge is attributed to an insufficient amount of hydroxy complex required for flotation. The pH ranges in which flotation is effected in the presence of a number of metal salts are listed in Table 7-3. These ranges are those in which flotation recovery greater than 90 percent is obtained. Regions of flotation response of quartz similar to those in Table 7-3 have been presented.

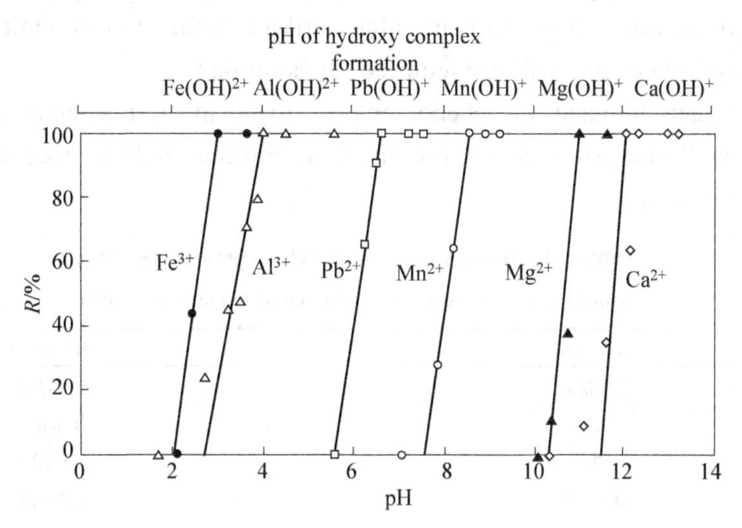

Figure 7-35 Minimum flotation edges of quartz as a function of pH

Conditions: 1×10^{-4} mol/L sulfonate, 1×10^{-4} mol/L metal ion

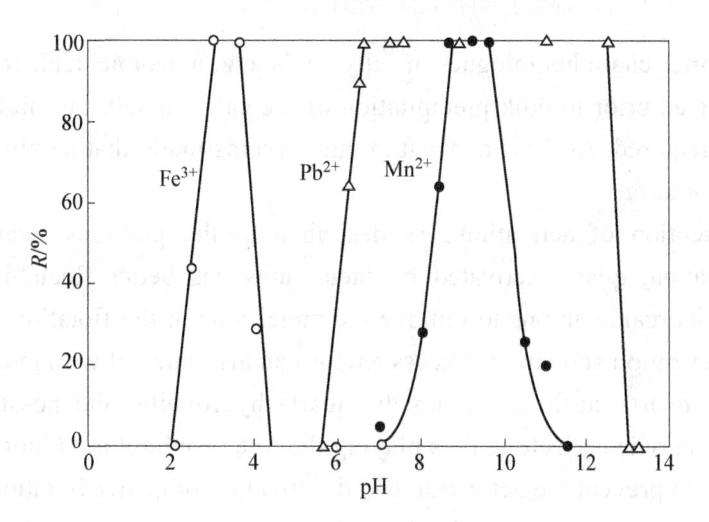

Figure 7-36 Flotation recovery of quartz as a function of pH

Conditions: 1×10^{-4} mol/L sulfonate, 1×10^{-4} mol/L metal ion

Similar behavior has been found for carboxylates. Flotation response in the presence of various additions of calcium chloride and lauric acid at pH 11.5 is presented in Figure 7-37. Arrows indicating the activity of laureate at which calcium laurate precipitates in each of the systems are shown. As can be noted, flotation occurs only after precipitation of calcium laureate has occurred. Similar results with oleate as collector have been reported.

The critical nature of the addition of collector relative to that of activator has been shown by some scholars. In those systems in which precipitation of metal collector has occurred, at constant collector addition, increasing the metal ion addition by an order of magnitude reduces the minimum pH at which flotation is possible by one unit.

On the other hand, at constant metal ion addition, increasing the collector addition reduces the pH at which flotation is possible until a critical level of addition is reached, at which point higher values of pH are required for flotation.

As these facts indicate, the phenomena occurring in these systems are complex, and they are not well understood at the present time. Further study is needed to delineate the mechanisms involved.

Table 7-3 Ranges in pH in which 90 percent recovery of quartz is obtained in the presence of various activators①

Metal salt	pH range
Fe(Ⅲ)	2.9-3.8
Al	3.8-8.4
Pb	6.5-12.0
Mn(Ⅱ)	8.5-9.4
Mg	10.9-11.7
Ca	12.0 and greater

① Conditions: 1×10^{-4} mol/L sulfonate, 1×10^{-4} mol/L metal ion.

With shorter-chain homologues of fatty acids, e.g. nonanoic acid, complete flotation of quartz is effected prior to bulk precipitation of the calcium salt, but high concentrations of collector are required. At these concentrations it seems likely that hemimicelles would have formed at the interface.

(2) Prevention of activation. As described in the previous section, using anionic collector flotation, quartz activated by metal ions has better floatability. Therefore, by adding some inorganic anions to remove the metal ions in the flotation system, it can play the role of prevention activation. Excess anions can also interact with metal salt ions already acting on the quartz surface, making the quartz hydrophilic and possibly preventing the adsorption of anionic collectors, thus playing the role of inhibitors. Fluoride and water glass are often used to prevent the activation and deactivation of quartz flotation.

Figure 7-38 shows the deactivation of quartz minerals activated by Al^{3+} by HF and Na_2SiO_3. Figure 7-38 shows that HF has a weak deactivation effect on quartz after Al^{3+} activation. Especially under the condition of near neutral pH around 7, there is no deactivation effect on quartz. This is consistent with the medium pH where HF has the best activation effect on quartz. Under the condition of pH<4, Na_2SiO_3 can activate quartz. Under the condition of pH>4, the quartz mineral activated by Al^{3+} has a good deactivation. When pH>6, Na_2SiO_3 can reduce the flotation recovery of quartz to less than 10%.

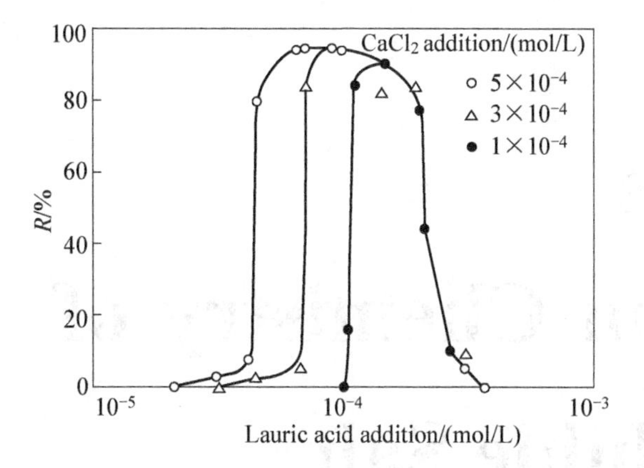

Figure 7-37 Flotation recovery of quartz as a function of lauric acid and calcium chloride additions at pH 11.5

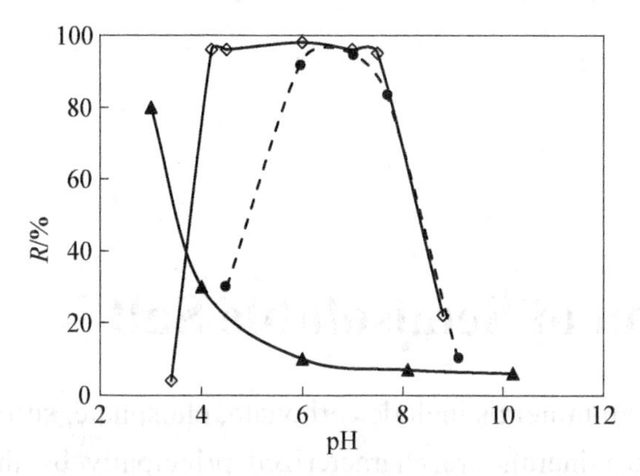

Figure 7-38 Flotation recovery of quartz activated by Al^{3+} as a function of pH in the absence and presence of HF or Na_2SiO_3

◇$AlCl_3$ 20 mg/L; ● $AlCl_3$ 20 mg/L+HF 40 mg/L; ▲$AlCl_3$ 20 mg/L+Na_2SiO_3 80 mg/L

Figure 7-39 shows the deactivation of Na_2SiO_3 on quartz minerals activated by Fe^{3+}. It can be seen from the figure that although the overall deactivation of Na_2SiO_3 on quartz is strong, there is basically no deactivation of Na_2SiF_6 when pH=5-6.5.

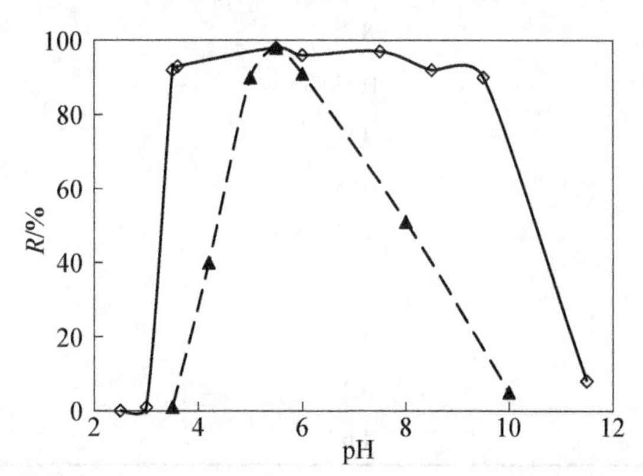

Figure 7-39 Flotation recovery of quartz activated by Fe^{3+} as a function of pH in the presence of $NaSiF_6$

◇ $FeCl_3$ 5×10^{-5} mol/L; ▲$FeCl_3$ 5×10^{-5} mol/L + Na_2SiF_6 4.25×10^{-5} mol/L

Chapter 8

Flotation Chemistry of Semisoluble Salt

8.1 Flotation of Semisoluble Salt

The semisoluble salt minerals include carbonate, phosphate, sulfate, tungstate and some halide minerals. These minerals are characterized principally by their ionic bonding and their moderate solubility in water. See Table 8-1. Many of these minerals have solubilities on the order of 10^{-4} mol/L.

Table 8-1 Solubility product constants and electrokinetic properties of various semisoluble salt minerals

Mineral	Solubility pK_{so}	Electrokinetic properties pzc or IEP (pH)②
Calcite	8.4	IEP 10.8
Aragonite	8.2	pzc$<$5
Dolomite	16.7	IEP$<$7
Magnesite	4.9	pzc 6-6.5
Fluorite①	10.3	IEP 10
Fluorapatite	—	IEP 5.6
Hydroxyapatite	—	pzc 8.5±0.2
Barite	9.7	pzc 3.4
Celestite	6.2	pzc 2.3
Gypsum	4.6	pzc 2.3-10.7

① Open to air.

② The isoelectric point of a solid is the point at which the zeta potential is zero. When the zeta potential is measured in the presence of uni-univalent acids or bases, the IEP and pzc of the solid are the same.

It would be expected that the surface charge of these minerals in water would be a function of the concentration of the ions of which their lattices are composed. This has been observed experimentally in some systems, and selected data are presented in Table 8-1. More frequently, however, the pzc of many of these minerals is determined by the hydrogen ion concentration. As discussed in the section on Flotation Theory, this dependency arises when the surface anion of the salt acts as a weak acid. Note, for example, that the pzc of calcite (pH=10.8) corresponds to the second dissociation constant of carbonic acid.

$$H_2CO_3 \rightleftharpoons H^+ + HCO_3^- \qquad K_1=4.16\times10^{-7} \tag{8-1}$$

$$HCO_3^- \rightleftharpoons H^+ + CO_3^{2-} \qquad K_2=4.84\times10^{-11} \tag{8-2}$$

As demonstrated in the case of oxides, the point-of-zero-charge (pzc) for semisoluble salts can also be estimated from solubility data, for saturated solutions as shown for the calcite system in Figure 8-1. Notice that the solution isoelectric point (IEP) calculated from thermodynamic data is 8.2, whereas the pzc of calcite is pH 10.8. However, after, equilibrium has been attained (after days of aging), the pzc is reported to approach pH 8.2, the value predicted from solubility calculations.

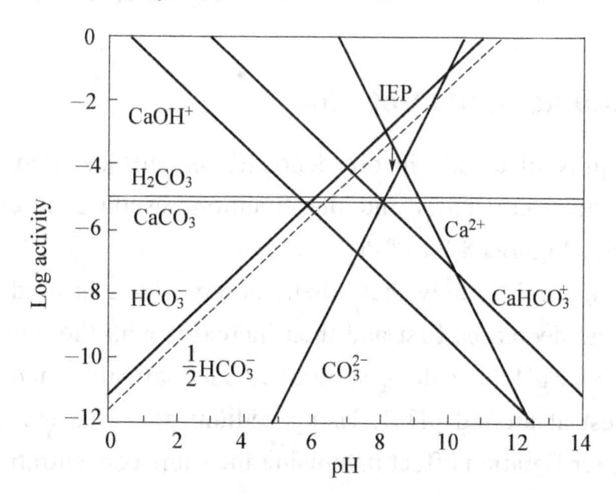

Figure 8-1 Determination of point-of-zero-charge of calcite from thermodynamic data

Another example where the surface anion acts as a weak acid is the apatite system, pzc pH 5-8. Again, as in the calcite system, the pzc is close to the second dissociation constant of phosphoric acid which has a pK_2 of 7.2.

$$H_3PO_4 \rightleftharpoons H^+ + H_2PO_4^- \qquad K_1=7.5\times10^{-3} \tag{8-3}$$

$$H_2PO_4^- \rightleftharpoons H^+ + HPO_4^{2-} \qquad K_2=6.2\times10^{-8} \tag{8-4}$$

$$HPO_4^{2-} \rightleftharpoons H^+ + PO_4^{3-} \qquad K_3=1.0\times10^{-11} \tag{8-5}$$

Anionic collectors are most frequently used in the flotation of semisoluble salt minerals. In particular, carboxylic acids are used extensively—unsaturated alkyl fatty acids and resin acids—for the alkaline earth minerals, and shorter chain (C_8) saturated alkyl fatty acids—

coconut all derivatives—for the metallic semisoluble salt minerals. Also of importance in the flotation of metallic semisoluble salt minerals are the sulfhydryl type collectors, such as the longer chain xanthates. In many instances their use requires sulfurization or else collector consumption becomes prohibitive. Other collectors, e.g. sulfonates, have rather limited industrial use in these systems.

In most systems it is difficult to achieve high selectivity, and desliming may be required as is the case in phosphate flotation and in many of the metallic systems. Flotation is generally accomplished in alkaline media. Soda ash rather than lime is used almost exclusively for pH control in order to avoid calcium activation of gangue silicates and also to avoid bulk phase precipitation of the calcium salt of the collector.

In most instances collector adsorption in these systems involves chemisorption. This phenomenon results from the stability of most multivalent cation carboxylate salts and the moderate solubility of the semisoluble salt minerals. Chemisorption has been inferred in a number of these systems both from direct and indirect evidence.

Infrared spectroscopy has been invaluable in the study of collector adsorption, particularly carboxylate adsorption because of scanning adsorption of infrared radiation by the carbonyl band.

8.1.1 Calcite and dolomite flotation

Due to the ubiquity of calcite in ore deposits, the surface properties, the adsorption characteristics of various collectors and the flotation response of calcite have received considerable study. See Figures 8-2 to 8-5.

The results in Figure 8-2 show that when sodium oleate is used as the collector, the recovery rate of calcite decreases first and then increases with the increase of pH value. In the range of pH 5.2 and pH 11, calcite showed good floatability, and the recovery rate of calcite was the lowest at around pH 7. Using sodium oleate as the collector to flotation calcite can obtain better flotation effect than using the same concentration of oleic acid.

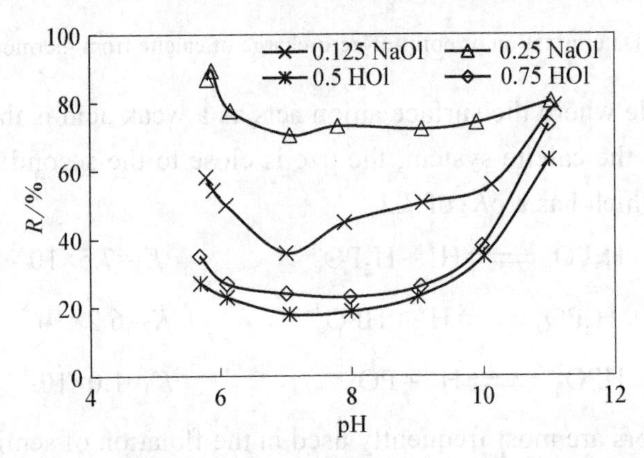

Figure 8-2 Flotation recovery of calcite at various pH and different concentrations of sodium oleate and oleic acid

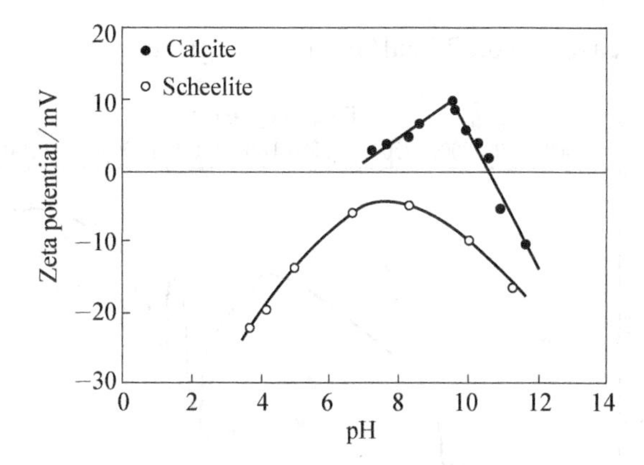

Figure 8-3 Zeta potential of calcite and scheelite as a function of pH

The nature of oleate adsorption on calcite is revealed partially by consideration of infrared spectra. The infrared patterns of calcite and calcium oleic, are presented in Figures 8-4 and 8-5. The deep absorption bands at 1543 and 1577 cm^{-1} represent the carbonyl stretching mode for calcium oleate. The chemisorption reaction which occurs at the calcite surface is revealed by the carbonyl absorption band which has shifted to 1587 cm^{-1} in the adsorbed state as shown in the differential spectra in Figure 8-5. There is no evidence of the physically adsorbed oleic acid molecule which would have an absorption band at 1718 cm^{-1}. In addition, infrared studies have proved that short chain carboxylates, namely lauric acid, are also chemisorption in calcite. With regard to flotation, recovery has been established as a function of collector concentration and hydrocarbon chain length.

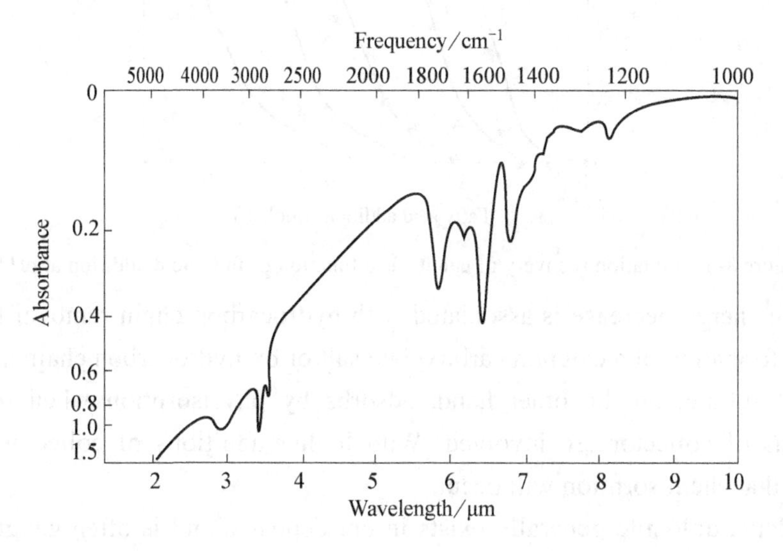

Figure 8-4 Infrared spectrum of calcium oleate adsorption at pH 5.5, with 2.0% KBr

Both fatty acids and sulfonates-sulfonates have been investigated, and the result obtained with fatty acids are presented in Figure 8-6. When the length of the hydrocarbon chain is increased, the concentration of carboxylate, necessary for flotation is reduced. A kT plot of these data yields a slope which corresponds to a specific adsorption potential of −1.1

kT or a free energy decrease of –0.62 kcal/mol of CH_2 group.

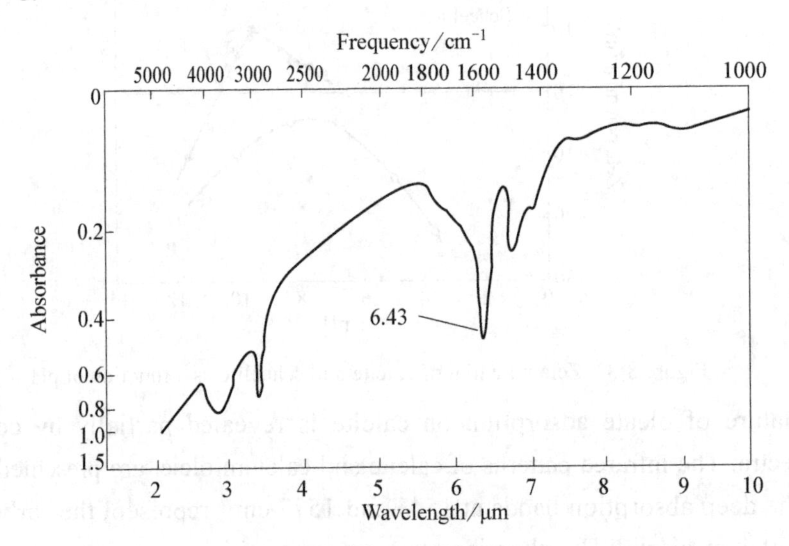

Figure 8-5 Infrared spectrum of chemisorbed oleate on calcite

(pH 5.5 adsorption, alcohol wash, 2.0% KBr)

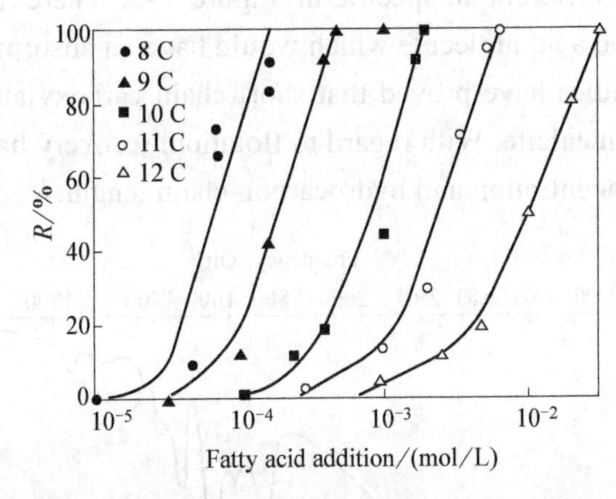

Figure 8-6 Flotation recovery of calcite as a function of fatty acid addition at pH 9.7

This free energy decrease is associated with hydrocarbon chain removal from solution either by the formation of a calcium carboxylate salt or by hydrocarbon chain association.

Dodecyl sulfate, on the other hand, adsorbs by physisorption when relatively low concentrations of collector are involved. With higher additions of collector, however, it seems likely that chemisorption will occur.

Like calcite, dolomite generally exists in ore deposits, and is often gangue mineral in flotation. Figure 8-7 shows the flotation recovery of dolomite as a function of pH at different concentrations of sodium oleate and oleic acid. Figure 8-7 shows that when sodium oleate is used as the collector, with the increase of pH value, the recovery of dolomite first decreases and then increases, showing a V-shaped change. When the pH value is around 7, the floatability of dolomite is the worst. The recovery rate of flotation with oleic acid as the collector is lower than that with sodium oleate. Under the reagent conditions in the figure, when the pH value is 6-10, the recovery rate of dolomite does not exceed 20%.

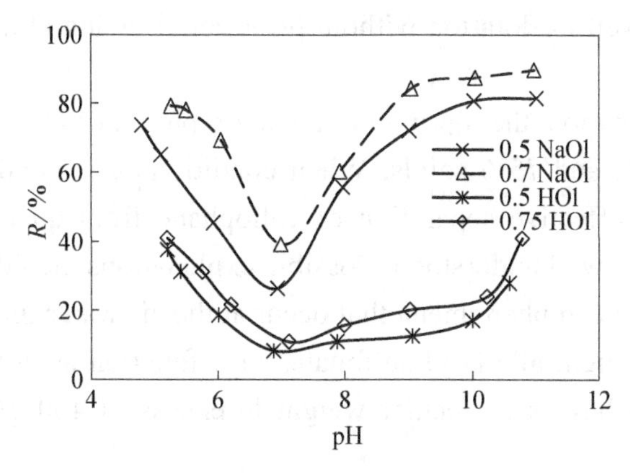

Figure 8-7 Flotation recovery of dolomite as a function of pH at different concentrations of sodium oleate and oleic acid

8.1.2 Apatite - collophane flotation

Calcium phosphate occurs as either the mineral apatite or as collophane, a substituted cryptocrystalline calcium phosphate, chemically similar to apatite but quite dissimilar with regard to appearance, surface area and porosity .

Electrokinetic experiments have shown that apatite has a point-of-zero-charge at pH 6.4. The complete flotation response from pH 5 to pH 13, shown in Figure 8-8, reveals that oleate is adsorbing chemically on the surface. For a negatively charged ion ($RCOO^-$) to adsorb on a negatively charged surface many units in pH above the pzc, a chemical reaction between carboxylate ion and surface calcium must be occurring.

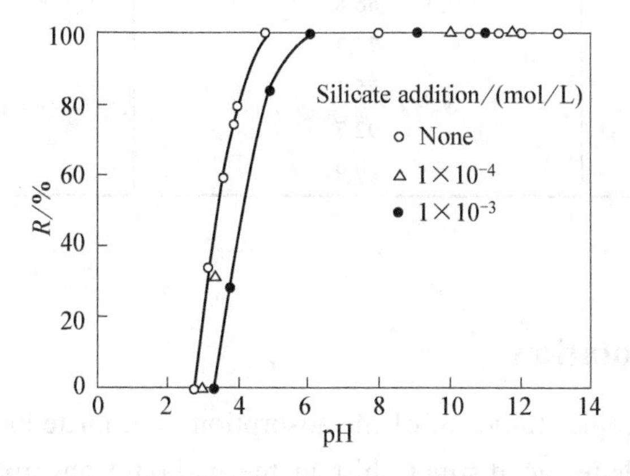

Figure 8-8 Flotation recovery of apatite as a function of pH with 1×10^{-4}mol/L oleate in the absence and presence of sodium silicate (Type N)

Processing of phosphate rock, containing nodules of collophane, involves desliming at 150 mesh. The minus 150-mesh material, which frequently contains one third of the P_2O_5 values, is discarded. Techniques features are involved in collophane flotation, namely the use of neutral hydrocarbon oils and conditioning at high pulp density. Both of these features probably reflect the porosity and resultant high surface area of collophane. Apatite, on the other hand, responds well to flotation without prior conditioning at high pulp density in the absence of fuel oil.

In collophane flotation the optimum range of pulp density during conditioning is generally between 70% and 75% solids. When conditioning is conducted such below this range in pulp density effective separation of collophane from quartz cannot be obtained. This phenomenon is not well understood. Possible explanations include the ratio of adsorbed collector to surface area and phenomena that occur at the air/water interface.

High-molecular weight alkylaryl sulfonates also function as collectors in this system, but the sulfonate must have a molecular weight in excess of 400 g/mol to effect a strong flotation response.

This work also demonstrates the critical nature of oil addition in collophane flotation in which oil of relatively high viscosity is required. See Table 8-2. Apparently, the function of the neutral hydrocarbon oil, whether the collector be a fatty acid or a sulfonate, is to bridge adsorbed collector molecules and to fill the pores of the highly porous cellophane. Such action would reduce collector consumption and aid in the surface dehydration process. Larger and more highly branched molecules would be required to accomplish this, perhaps explaining the effect of viscosity of the oil.

Table 8-2 Effect of oil viscosity on phosphate recovery at pH 7.5, 0.27 lb/t sulfonate (molecular weight 450-470) and a pulp density of 71% solids

Oil viscosity (sec. Saybolt at 100℉)	Rougher phosphate recovery /%	Cleaner phosphate recovery /%
No oil	68.8	1.8
32.5 (0.40 lb/t)	75.3	1.8
33.8 (0.36 lb/t)	75.1	1.1
378 (0.46 lb/t)	92.7	8.1
407 (0.46 lb/t)	89.9	9.9

Note: $t/℃=\frac{5}{9}(t/℉-32)$.

8.1.3 Fluorite flotation

Infrared spectroscopic studies of oleate adsorption on fluorite have been conducted, but as contrasted with calcite, no distinct shift in the carbonyl absorption band is observed. Other researchers have reported similar spectra which substantiate the presence of a surface calcium oleate species. That is, the pzc of fluorite is pH 10.0 and complete flotation is effected even at approximately pH 13. See Figure 8-9. With a zeta potential of approximately –35 mV at pH 11, adsorption of oleate ion could occur only by chemical reaction with surface calcium ion. Further, the stoichiometric release of fluoride ions with oleate adsorption was observed.

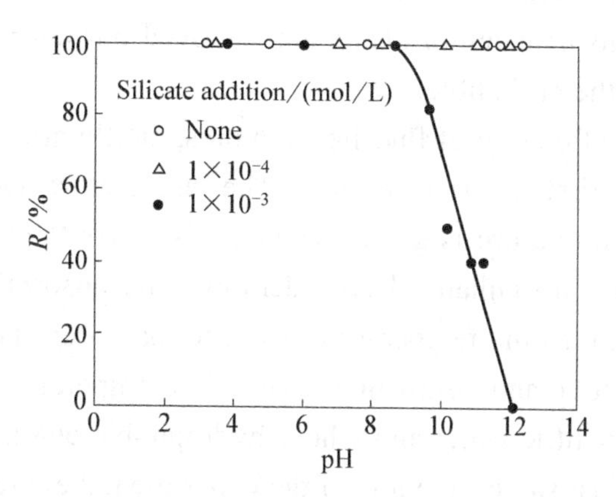

Figure 8-9 Flotation recovery of fluorite as a function of pH with 1×10^{-4} mol/L oleate in the absence and presence of Type N sodium silicate

The distribution of adsorbed oleate species as a function of pH has been determined by infrared spectroscopy and is presented in Figure 8-10. Notice that the predominate surface species in the region of optimum flotation is chemisorbed calcium oleate. Detailed solution chemistry analysis of the fluorite and other systems as a function of pH supports these data. The stability region for HOl and $Ca(Ol)_{2(S)}$ are superimposed on the distribution diagram. These results again reinforce the premise that in oleate flotation of nonsulfide minerals, the adsorption process involves a chemical reaction leading to the formation of multilayers of the collector soap.

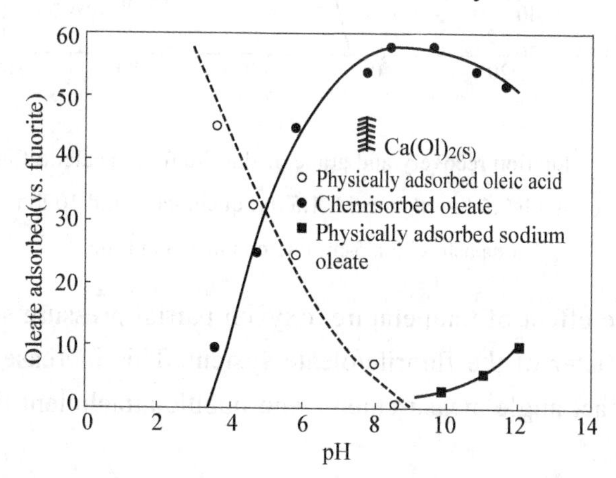

Figure 8-10 Surface associated oleate on fluorite as a function of pH

The added oleate concentration was 9×10^{-4} mol/L. The different surface species were identified from the infrared spectra of the fluorite-oleate aggregate. The hachured boundary, derived from the stability constants, denotes the location of the bulk stability limit for oleate species

Studies of the fluorite/oleate system provide evidence that the adsorption of multiplayer is a chemical reaction and, in fact, calcium oleate has been identified as a separate phase. The relative proportion of tightly-held chemisorbed calcium oleate multilayers to loosely-held colloidal calcium oleate precipitate has been reported to be significantly higher for fluorite than for calcite. Such a phenomenon helps to explain the difference in the floatability of these minerals. Further, it has been established that under certain circumstances these films of calcium oleate are removed during bubble attachment simply by the buoyant force of the air bubble.

In the case of oleate flotation of fluorite from ores, the flotation response is enhanced at elevated temperature as first reported in 1925. The temperature sensitivity of the flotation response of fluorite from one ore is given in Figure 8-11. As the temperature is increased, both grade and recovery are enhanced considerably. The adsorption mechanism changes from physical to chemical as the temperature is raised, based on the physical differences in concentrates obtained above and below 60℃. The high temperature concentrate was a very dry froth containing very little water and whose hydrophobic character did not change even with repeated cleanings. However, in view of the experimental evidence which indicates that chemisorptions of oleate occur at room temperature, it would appear that phenomena other than chemisorption of oleate are occurring when elevated temperature is involved.

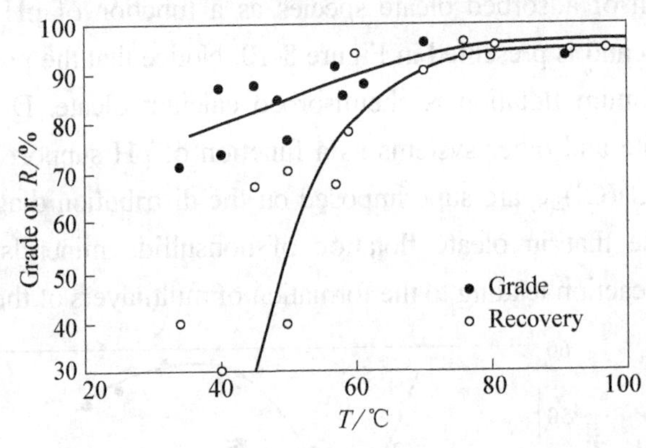

Figure 8-11 Flotation recovery and grade of fluorite from a calcite-fluorite-barite ore with 2.5 lb/t oleic acid; 1.5 lb/t quebracho; and 10 lb/t soda ash with a 5-minute conditioning time

In addition to the effect of temperature, oxygen partial pressure significantly increases the hydrophobic character of the fluorite/oleate system. This increase in hydrophobicity is revealed by both contact angle measurements and bubble attachment time measurements as illustrated in Table 8-3.

Table 8-3 Effect of oxygen on bubble attachment

Gas phase	Contact angle/(°)	Attachment time/ms
Air (22℃)	70-73	260-270
Oxygen(22℃)	90-92	20-25

Note: Measurements at a fluorite surface, conditioning time ~20 minutes, oleate concentration 10^{-5} mol/L, pH 8.1, bubble size ~3.8 mm.

This increase in hydrophobicity is similar to the increase observed at elevated temperature. When the ineffectiveness of saturated fatty acids as collectors is considered, these results suggest that double bond interaction may be an important factor in oleate flotation systems. Some evidence suggests that cross linking may occur between adsorbed oleyl chains via an epoxide, linkage leading to polymerized surface species with greater hydrophobic character.

8.1.4 Anglesite, cerussite and malachite flotation

Anglesite, cerussite and malachite all respond well to flotation with relatively long-chain fatty acids due to the insoluble nature of heavy-metal soaps and also to the hydrophobicity imparted by long-chain collectors. When short-chain xanthates are used as collector, however, these minerals are not floated nearly as effectively as their sulfide counterparts. This fact is due to the extensive hydration of the carbonate and sulfate surfaces that occurs in these systems as compared to sulfide minerals under similar circumstances.

As shown in Figure 8-12, both anglesite and cerussite are floated completely with amyl xanthate. In the case of anglesite, a ten-fold increase in collector concentration over that required to float cerussite is needed. This phenomenon is related to the solubility of each of these minerals, that is approximately 10^{-4} mol/L Pb^{2+} will be contributed to solution from anglesite whereas approximately 10^{-6} mol/L Pb^{2+} will be in equilibrium with cerussite.

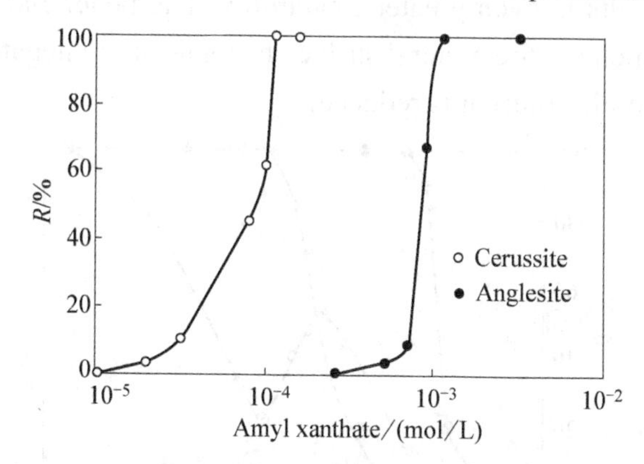

Figure 8-12 Flotation recovery of anglesite and cerussite as a function of amyl xanthate concentration at pH 9.6

The benefit of sulfurization of these minerals with sodium sulfide has been known for some time. After sulfurization the surface is much less hydrophilic due to the presence of chemisorbed sulfide ion. Under these conditions collector requirements are reduced significantly in magnitude.

The critical nature of sulfide ion addition in sulfurizing anglesite is illustrated graphically in Figure 8-13. Flotation is possible only in a narrow rang of sulfide addition. Below 4.2×10^{-4} mol/L sulfide, S^{2-} is precipitated by dissolved Pb^{2+} as PbS. After essentially all of the Pb^{2+} has been removed from solution, S^{2-} adsorbs on the anglesite surface forming a surface lead sulfide, and flotation is achieved under these conditions.

When additions of sulfide above 5.3×10^{-4} mol/L are made, however, depression is again experienced due to sulfide ion adsorption in preference to xanthate adsorption.

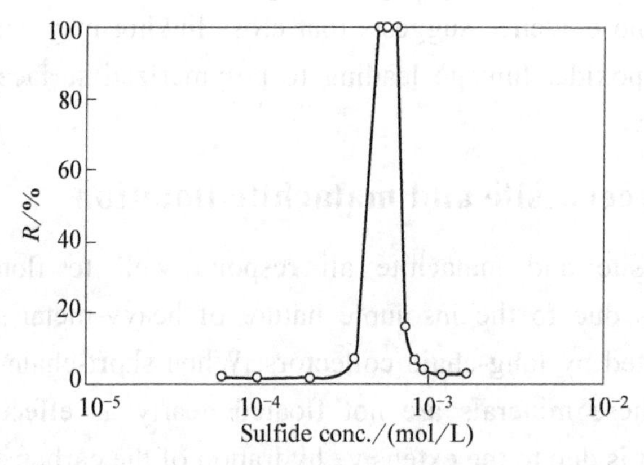

Figure 8-13 Flotation recovery of anglesite as a function of sulfide ion addition with 1×10^{-4} mol/L xanthate at pH 9.6

Amines also float these minerals, more effectively in the presence of sodium carbonate and after sulfurization, however. High-molecular weight saturated amines (60% C_{18}, 30% C_{16}) are generally used as collector. As can be noted in Figure 8-14, the addition of sodium carbonate increases floatability of anglesite, while the addition of both sodium carbonate and sodium sulfide results in even greater floatability. The beneficial effect of carbonate is probably two folds, which are the mineral surface becomes more negatively charged, and the concentration of Pb^{2+} ion in solution is reduced.

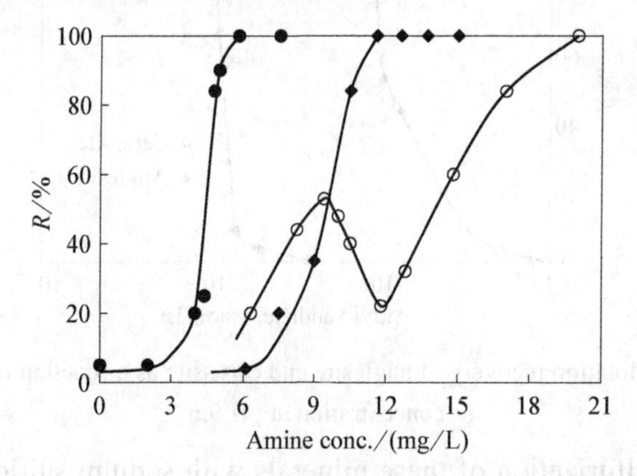

Figure 8-14 Flotation recovery of anglesite as a function of amine concentration at pH 8.6 in the absence and presence of ammonium sulfide and sodium bicarbonate

Reagent additions: ○ None; ● 1×10^{-3} mol/L HCO_3^-; ◆ 1×10^{-3} mol/L HCO_3^-, 5×10^{-4} mol/L S^{2-}

The enhanced floatability obtained after sulfurization is probably due to the increased hydrophobicity of the surface under these conditions.

8.2 Modifiers and Depressants

Mechanisms involved in depression reactions have not been investigated thoroughly, and as a result, limited understanding of these systems has been achieved. Not only are the depressant adsorption reactions complex, but the depressants are frequently complex molecules themselves. The basic depressants used in semisoluble salt flotation systems are: sodium carbonate, sodium silicate, sodium fluorosilicate, sodium metaphosphate, lignin sulfonate, humic acid, starch, quebracho and tannin derivatives.

8.2.1 Inorganic modifiers and depressants

(1) Sodium carbonate. Surface carbonation is an important reaction that occurs in semisoluble salt flotation systems, resulting from CO_3^{2-} in solution derived from the CO_2 in the atmosphere or from carbonate minerals present.

In the case of fluorite, dissolved carbonate reacts with the surface to form a calcium carbonate compound. As a matter of fact, this phenomenon accounts for the pzc of fluorite being at pH 10.0. Calculation shows that this reaction should occur at pH values greater than pH 8 for a system open to the atmosphere. As shown in Figure 8-15, the infrared spectra for fluorite conditioned at pH values greater than pH 8 and open to the atmosphere are characterized by a doublet with absorption band at 1400 cm^{-1} and 1480 cm^{-1}. Doublets such as this suggest that the surface carbonate is unsymmetrical and in unidentate (one oxygen of carbonate adsorbed) coordination with the surface. A symmetrical carbonate ion in which the carbon-oxygen bonds all vibrate at the same frequency is characterized by a single absorption band at 1400 cm^{-1}, whereas a doublet, such as is observed in this system, suggests that the surface carbonate is unsymmetrical and in unidentate coordination with the surface.

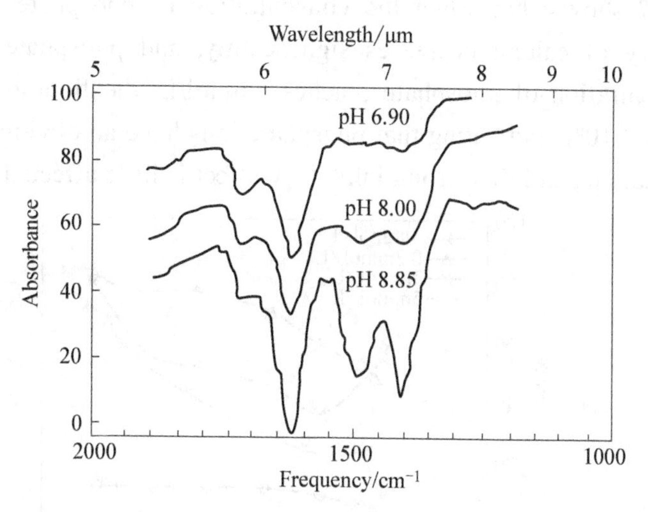

Figure 8-15 Infrared spectra of fluorite conditioned at various values of pH in systems open to air

In another study, examination of the infrared spectra of barite prepared in an aqueous suspension reveals the same doublet observed with fluorite samples, whereas spectra of barite prepared in non-aqueous environments give no indication of surface carbonation.

The practical implication of these phenomena is that in the flotation of semisoluble salts where soda ash is used for pH control, surface carbonation seems inevitable, and the fact that good selectivity is achieved seems remarkable.

(2) Sodium phosphate. Phosphoric acid, phosphate and their derivatives are often used as flotation modifiers for soluble minerals. Figures 8-16 to 8-19 show the flotation recoveries of fluorite, calcite, dolomite and collophane as a function of pH under different sodium phosphate concentrations.

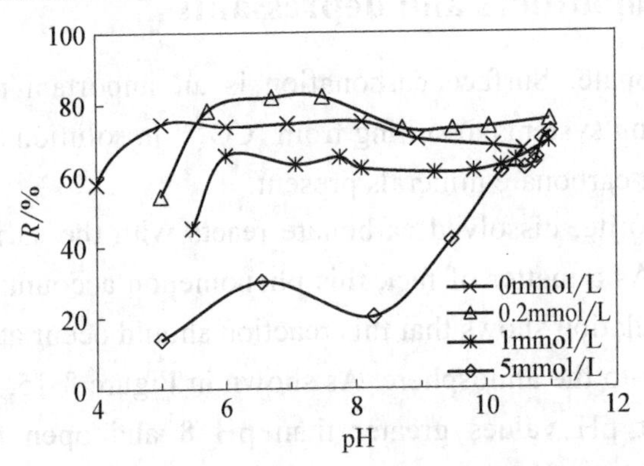

Figure 8-16 Flotation recovery of fluorite as a function of pH with different sodium phosphate concentrations

As can be seen from Figure 8-16, when the amount of sodium phosphate is 5 mmol and the pH is less than about 10.5, the floatability of fluorite decreases significantly, especially under acidic conditions. Phosphate had a strong inhibitory effect on fluorite, and the inhibitory effect increased with the increase of phosphate concentration.

Figure 8-17 shows that when the concentration of phosphate ion is 1mmol/L, the flotation recovery of calcite decreases significantly, and phosphate ion inhibits calcite. When the concentration of phosphate reaches 5mmol/L, the flotation recovery of calcite decreases to about 10%, indicating that phosphate ions have an obvious inhibitory effect on the flotation of calcite, and this strong inhibitory effect is little affected by pH value.

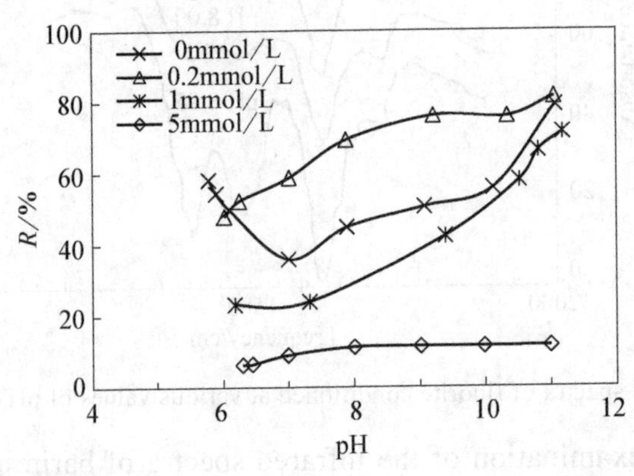

Figure 8-17 Flotation recovery of calcite as a function of pH with different sodium phosphate concentrations

Figure 8-18 shows that the added phosphate ion concentration and pH have a greater impact on dolomite flotation. When the phosphate ion concentration increased to 5 mmol/L, the flotation of dolomite was strongly inhibited. At this time, the flotation recovery rate of dolomite decreased with the decrease of pH.

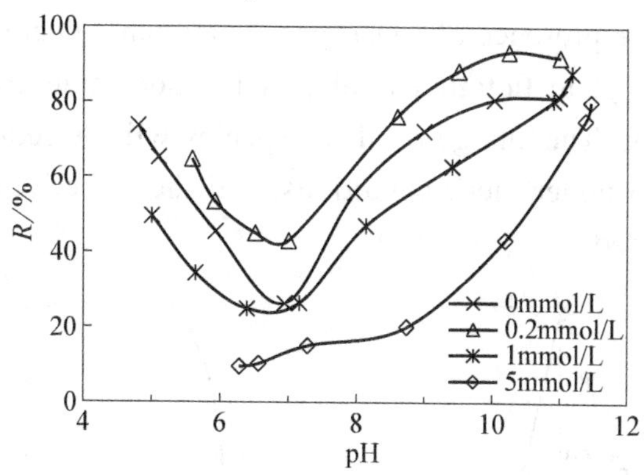

Figure 8-18 Flotation recovery of dolomite as a function of pH with different sodium phosphate concentrations

Figure 8-19 shows that when the added phosphoric acid concentration increases to 5 mmol/L, phosphate radical has a strong inhibitory effect on collophanite. When the pH value is less than 7.5, the recovery rate of collophanite is 0.

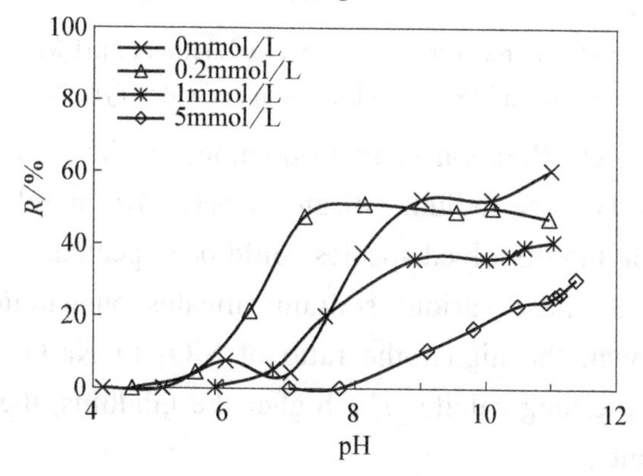

Figure 8-19 Flotation recovery of collophane as a function of pH with different sodium phosphate concentrations

High concentration of sodium phosphate has a strong inhibitory effect on these soluble minerals, while low concentration of phosphate ions has different inhibitory effects on these different minerals. Therefore, when these minerals are separated by flotation, sodium phosphate can be used as an inhibitor by taking advantage of the inhibition property of sodium phosphate.

(3) Sodium silicate. The composition of aqueous sodium silicates is expressed by the general formula, $m\text{Na}_2\text{O} \cdot n\text{SiO}_2$, where the ratio n/m is referred to as the modulus of sodium silicates. Commercial sodium silicates are available with ratios of SiO_2/Na_2O of 1.6, 2.75, 3.22 and 3.75. The commercial sodium silicate most widely used industrially as a dispersant or depressant is Type N (SiO_2/Na_2O-3.22).

The critical nature of the use of this reagent in achieving selectivity in nonmetallic flotation systems has been presented. In their study on apatite, calcite and fluorite, calcite is by far the most sensitive to sodium silicate additions. Flotation recovery of calcite with 5×10^{-4} mol/L oleate in the presence of 5×10^{-4} mol/L sodium silicate is given in Figure 8-20. At pH 6 essentially complete flotation is obtained. Under acid conditions protonation of silicate anions will occur. The charge on these species will be reduced (to perhaps even a neutral aqueous species) under these conditions, and the effectiveness of these species as depressants will be reduced.

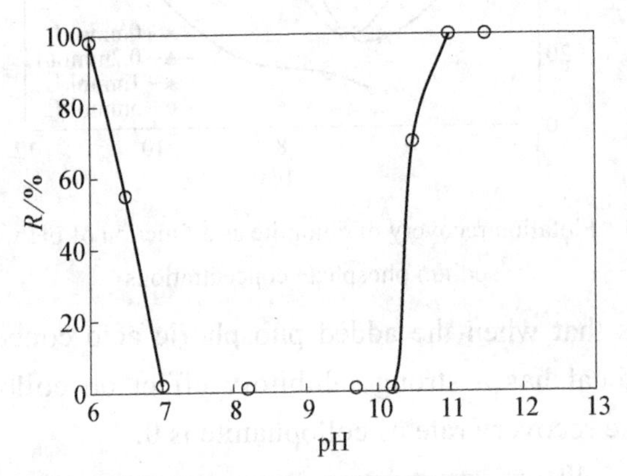

Figure 8-20 Flotation recovery of calcite as a function of pH with 5×10^{-4} mol/L oleate and 5×10^{-4} mol/L sodium silicate (Type N)

Above pH 10 complete flotation is again obtained in the presence of sodium silicate. The natural domain of Type N sodium silicate is pH 7-10, at relatively high pH values, different equilibria involving dissolved species could be expected.

The effectiveness of these various sodium silicates on calcite flotation is given in Figure 8-21. As is shown, the higher the ratio of SiO_2 to Na_2O, the more effective the sodium silicate is in depressing calcite. The higher the modulus, the greater is the quantity of silicate as a depressant.

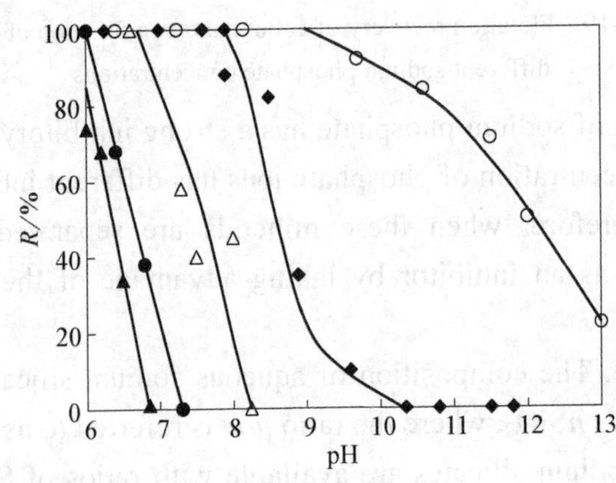

Figure 8-21 Flotation recovery of calcite as a function of pH with 1×10^{-4} mol/L oleate in the absence and presence of 5×10^{-4} mol/L commercial sodium silicates containing various ratios of SiO_2:Na_2O

○ None ; ◆ 1.60:1.0; △ 2.40:1.0; ● 3.22:1.0; ▲ 3.75:1.0

In another system, calcium silicate precipitates on the quartz and goethite surfaces are responsible for the dispersion.

In the case of fluorite, an addition of 1×10^{-4} mol/L sodium silicate has no effect on flotation response (Figure 8-9). With an addition of 1×10^{-3} mol/L, however, flotation recovery is reduced above about pH 10. With apatite the system is depressed about one-half unit in pH sooner in acid medium in the presence of 1×10^{-3} mol/L sodium silicate, but no effect is observed in basic medium (Figure 8-8). Flotation of scheelite under these conditions is presented in Figure 8-22. As can be noted, even 1×10^{-3} mol/L sodium silicate has no effect on flotation response.

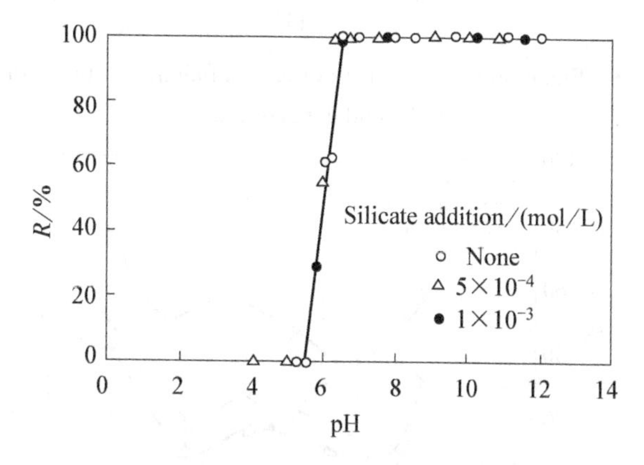

Figure 8-22 Flotation recovery of scheelite as a function of pH in the absence and presence of sodium silicate (Type N)

The variation in the strength of the surface calcium-oleate bonds on these minerals is apparent from these data. That is, the bond is apparently much weaker in the case of calcite than with the other three minerals in which silicate anions are able to compete effectively with oleate for surface sites and cause system depression.

8.2.2 Organic modifiers and depressants

(1) Humic acid. Humic acid is a kind of organic substance formed and accumulated through a series of processes of microbial decomposition and transformation and geochemistry. The basic structure of humic acid macromolecules is aromatic ring and aliphatic ring. The ring is connected with carboxyl, hydroxyl, carbonyl, quinone, methoxy and other functional groups.

Figure 8-23 to Figure 8-26 show the flotation recoveries of fluorite, calcite, dolomite and collophane as a function of pH under different humic acid concentrations.

Figure 8-23 shows that humic acid has little effect on fluorite flotation. When the added humic acid concentration is 0.5g/L and 1g/L, the fluorite flotation recovery rate is not much different from that without adding humic acid.

Figure 8-24 shows that humic acid has strong inhibition on calcite flotation. In the range of pH 6-10, the flotation recovery of calcite decreased from more than 40% to less than 20%. The flotation recovery decreased obviously with the increase of humic acid concentration.

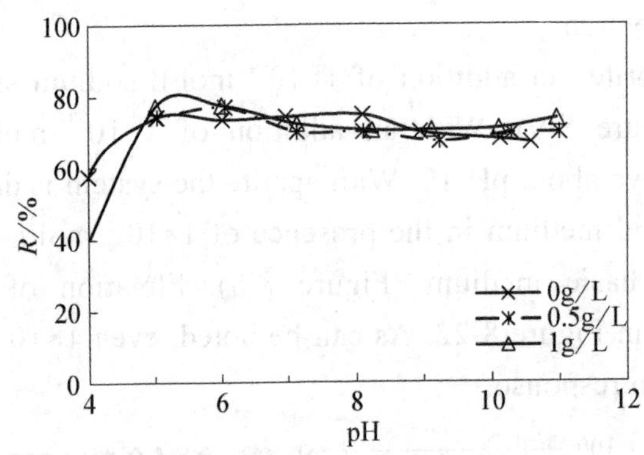

Figure 8-23 Flotation recovery of fluorite as a function of pH with different humic acid concentrations

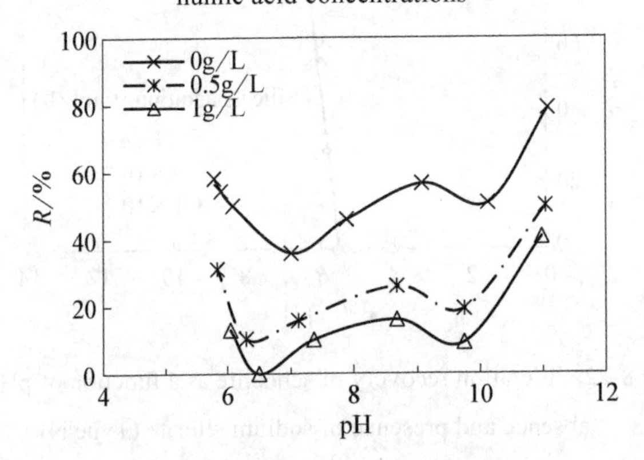

Figure 8-24 Flotation recovery of calcite as a function of pH with different humic acid concentrations

Humic acid has a strong inhibitory effect on dolomite flotation, which is more obvious under acidic and strongly alkaline conditions, as shown in Figure 8-25. When the humic acid concentration is 0.5g/L or 1g/L, the change of humic acid concentration has little effect on the flotation recovery of dolomite, and the flotation recovery basically does not exceed 20%. When the pH is 6-10, the flotation recovery is the lowest, is about 10%.

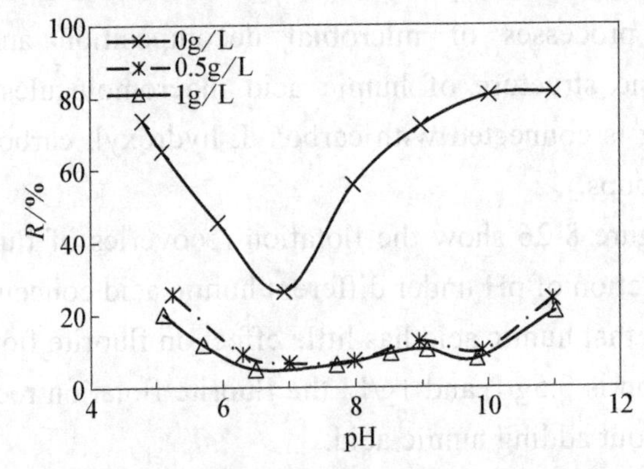

Figure 8-25 Flotation recovery of dolomite as a function of pH with different humic acid concentrations

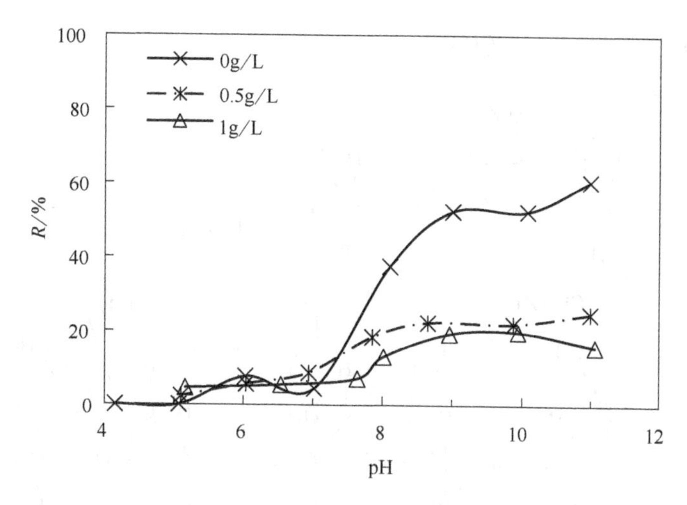

Figure 8-26 Flotation recovery of collophane as a function of pH with different humic acid concentrations

As can be seen from Figure 8-26, in the range of pH value greater than 8, the inhibitory effect of humic acid on the collophane is obvious, and the flotation recovery is reduced from more than 40% to less than 20%. The flotation depression of collophane was enhanced with increasing pH and humic acid concentration. When the pH is less than 7, the flotation recovery rate of collophane is less affected, and the flotation recovery rate is less than 10%.

It can be seen from the above that humic acid has a selective depression on calcite and dolomite, and using humic acid as depressant is beneficial to the flotation separation of fluorite from calcite and dolomite minerals.

(2) Starches and dextrose. Starch was recommended as a depressant as early as 1931. Industrial application has been extensive since that time, but basic systematic work dealing with mechanisms governing the depressant action has been limited.

The basic component of starch and dextrin is the dextrose molecule:

```
            OH
            |
       H—C6—H
            |
         HC5————O
        H/        \H
        |/         \|
       C4 OH     H  C1≡G
        |\       | /|
      OH|C3————C2  OH
          |      |
          H      OH
```

Dextrose

Structurally, starches and dextrin are polymers of dextrose monomeric units linked through 1-4 glycosidic linkages (for straight chain) and 1-6 linkages (for branch points). Straight-chain and branch-point linkages are as follows:

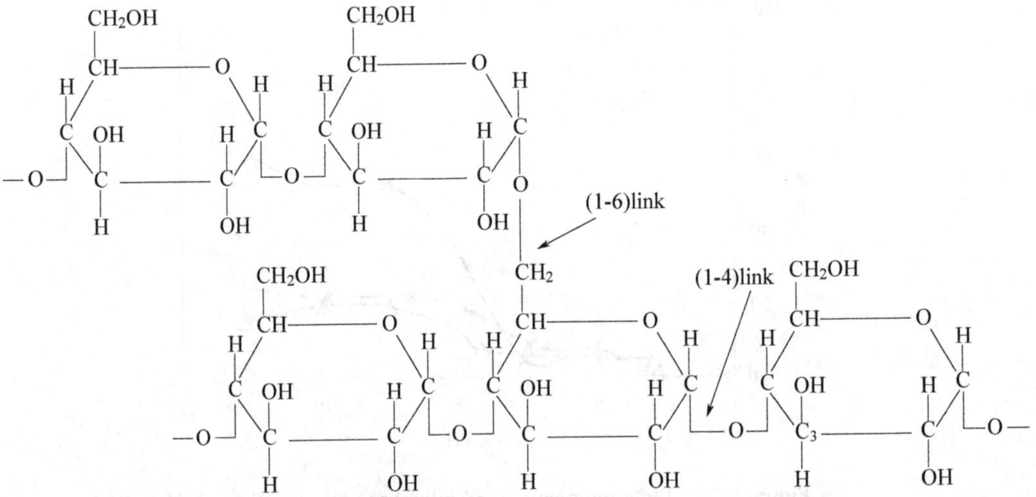

```
      CH2OH                       CH2OH
       |                           |
       CH———————O                  CH———————O
    H /          \ H            H /          \ H
     |/           \|             |/           \|
     C   OH      H  C            C   OH      H  C
 —O—┘ \ |          |/ └—O—┘ \ |          |/ |
       C ————————— C              C ————————— C   O
       |           |              |           |   |      (1-6)link
       H           OH             H           OH  |
                                                  CH2
            CH2OH                                 |            (1-4)link  CH2OH
             |                                    CH———————O                |
             CH———————O                        H /          \ H        H    CH———————O
          H /          \ H                      |/           \|        /            \ H
           C   OH      H  C                     C   OH      H  C      C   OH      H  C
     —O—┘ \ |          |/ └—O—┘ \ |          |/ └—O—┘ \ |          |/ └—O—
             C ————————— C                      C ————————— C         C3 ————————— C
             |           |                      |           |         |            |
             H           OH                     H           OH        H            OH
```

In starches the linear chain (amylose) and branched chain (amylopectin) components have molecular weights reaching millions. In dextrin formation these chains are fragmented and recombined to form low molecular weight but highly branched structures. Structural formulas of amylose, amylopectin and dextrin are:

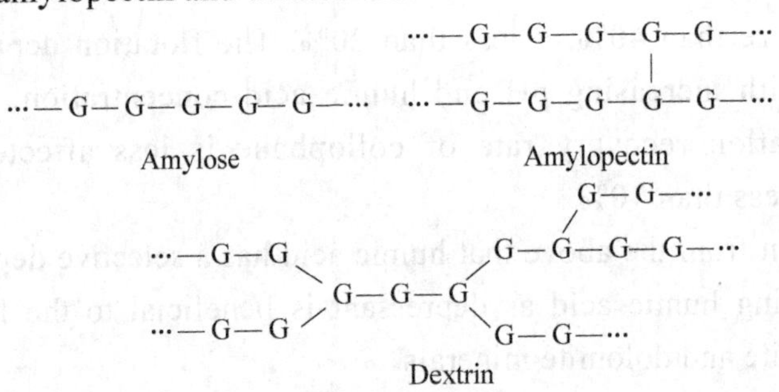

```
                                          ···—G—G—G—G—G—···
                                                      |
···—G—G—G—G—G—···                         ···—G—G—G—G—G—···
      Amylose                                  Amylopectin

                                     G—G—···
                                    /
           ···—G—G              G—G—G—G—···
                   \           /
                    G—G—G
                   /           \
           ···—G—G              G—G—···
                       Dextrin
```

Starches are anionic species, and their adsorption behavior is strongly affected by their molecular weight. Adsorption can occur by coulombic attraction and by hydrogen bonding with surface oxygen atoms.

In the few systems that have been studied in detail, coadsorption of collector with starch or starch derivatives has been reported. In the case of the calcite/oleate/starch system, the starch actually causes an increase in oleate adsorption. The calcite surface, however, becomes hydrophilic. It seems that such organic colloids can actually blind the hydrocarbon chain of the collector and project its polar hydroxyl groups, thus creating a hydrophilic surface. In the case of semisoluble salts, this is particularly true for calcite which may be related to the nature of the adsorbed calcium oleate at the calcite surface as discussed previously.

In a study of starch adsorption and its effect on semisoluble salt flotation with oleic acid, calcite is depressed at lower concentrations of starch than is barite which, in turn, is depressed at lower concentrations than fluorite. See Figure 8-27.

(3) Tannin. Tannins and their derivatives are also modifiers of semisoluble salt minerals. Structure of tannin is shown in Figure 8-28.

Quebracho, used to depress calcite, is a derivative of tannin. Its effectiveness has been well demonstrated in number of instances, but a complete understanding of its behavior has not been realized. In fact, a minimal amount of theoretical work on the flotation chemistry of this reagent has been done. The chemistry of quebracho itself has received considerable attention, and its complexity has been well established.

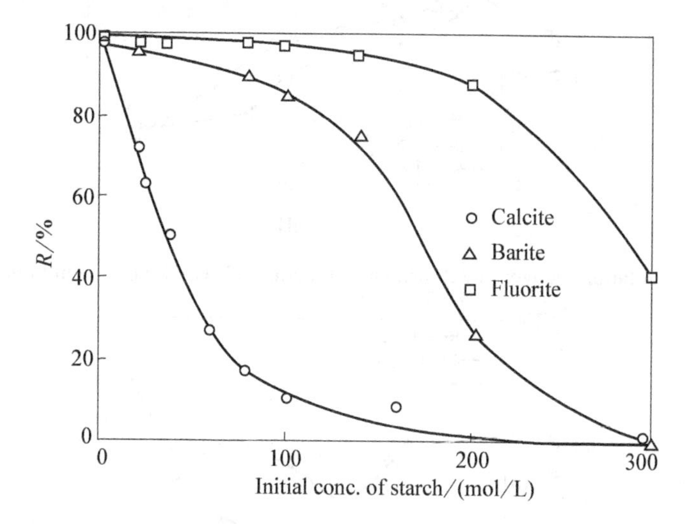

Figure 8-27 Flotation recovery of barite, calcite, and fluorite as a function of starch concentration with 7×10^{-5} mol/L oleic acid at pH 8

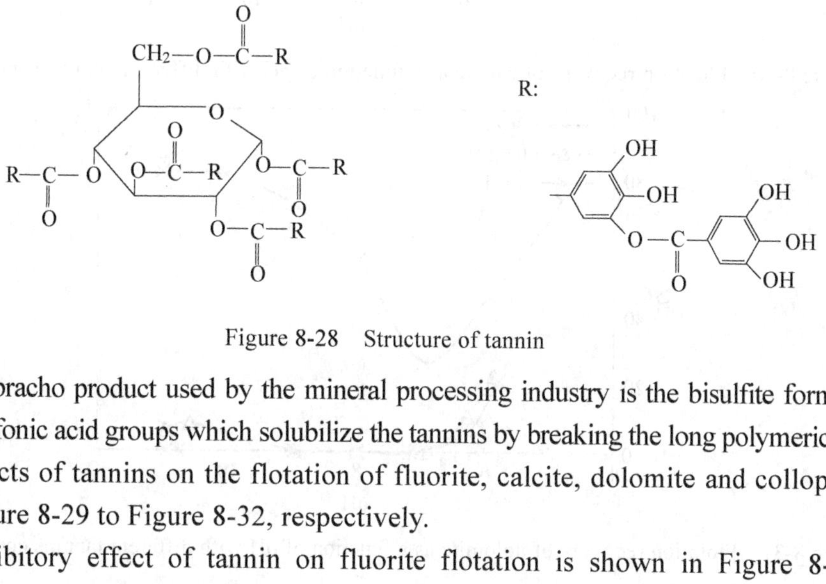

Figure 8-28 Structure of tannin

The quebracho product used by the mineral processing industry is the bisulfite form and has substituted sulfonic acid groups which solubilize the tannins by breaking the long polymeric chains.

The effects of tannins on the flotation of fluorite, calcite, dolomite and collophane are shown in Figure 8-29 to Figure 8-32, respectively.

The inhibitory effect of tannin on fluorite flotation is shown in Figure 8-29. The inhibition increased with the increase of pH and tannin concentration.

Figure 8-30 shows that under the inhibition of tannin, the flotation recovery of calcite is at a low point around pH 7, and the flotation recovery of calcite is lower than 20%. When the pH is greater than 9.5, calcite is strongly inhibited by tannin.

Figure 8-31 and Figure 8-32 show that dolomite flotation is strongly inhibited by tannin, and this inhibition is more intense under alkaline conditions. Tannins have a strong inhibition on the flotation of collophane. When the tannin concentration reaches 1 g/L, the flotation recovery of collophane is almost zero.

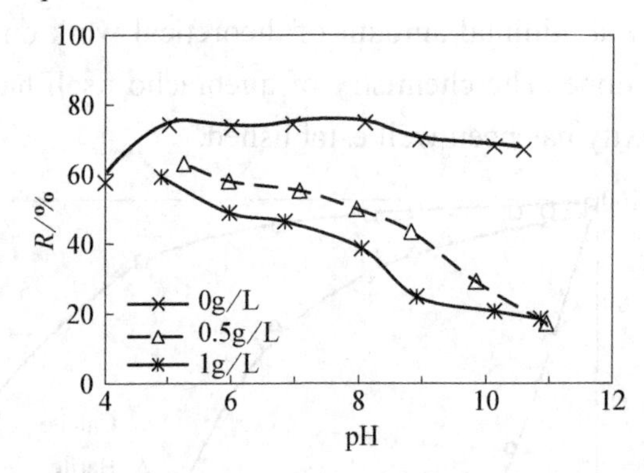

Figure 8-29 Flotation recovery of fluorite as a function of pH with different tannins concentrations

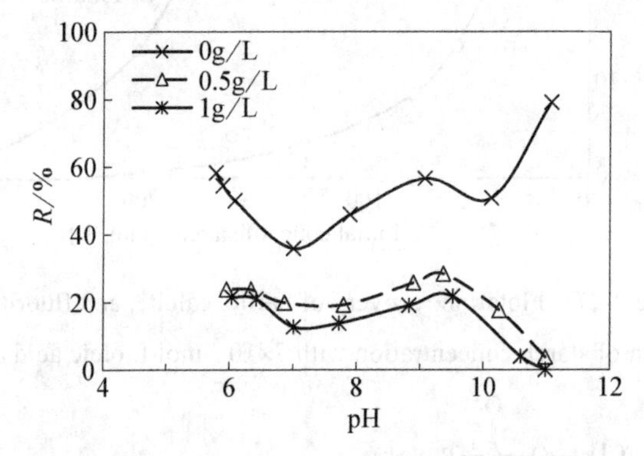

Figure 8-30 Flotation recovery of calcite as a function of pH with different tannins concentrations

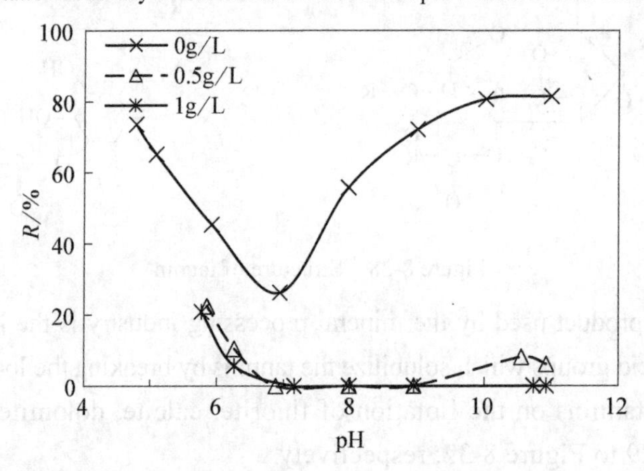

Figure 8-31 Flotation recovery of dolomite as a function of pH with different tannins concentrations

Comparison of the above test results shows that tannin has a strong inhibition on fluorite, calcite, dolomite and collophanite under the alkaline condition with pH value greater than 9. The inhibition of calcite is much stronger than fluorite in the range of pH 6-8.

Various mechanisms of tannin adsorption have been proposed. Some researchers attribute adsorption on calcite to electrostatic interaction between the positively charged mineral surface and the tannin species. On the other hand, some researchers attribute it to hydrogen bonding. A recent infrared study has provided evidence for the presence of calcium tannate complexes on the calcite surface.

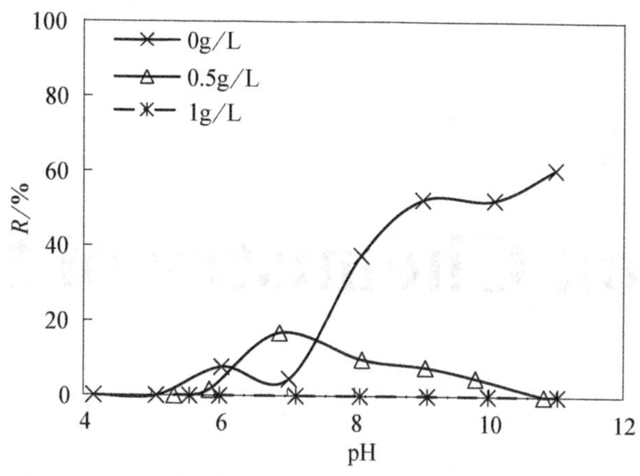

Figure 8-32 Flotation recovery of collophane as a function of pH with different tannins concentrations

The authors of this book believe that the adsorption of tannin on the calcite surface is shown in Figure 8-33. The dissolution of calcite results in the presence of exposed Ca lattice ions on the calcite surface. The polar groups carboxyl and hydroxyl in the organic molecular structure of humic acid sodium and tannin react with Ca by bonding or complexation. These organic substances are thus adsorbed on the mineral surface, while the polar groups such as hydroxyl and carboxyl groups on other unabsorbed ends are adsorbed with water molecules to make the minerals hydrophilic, which inhibits the calcite.

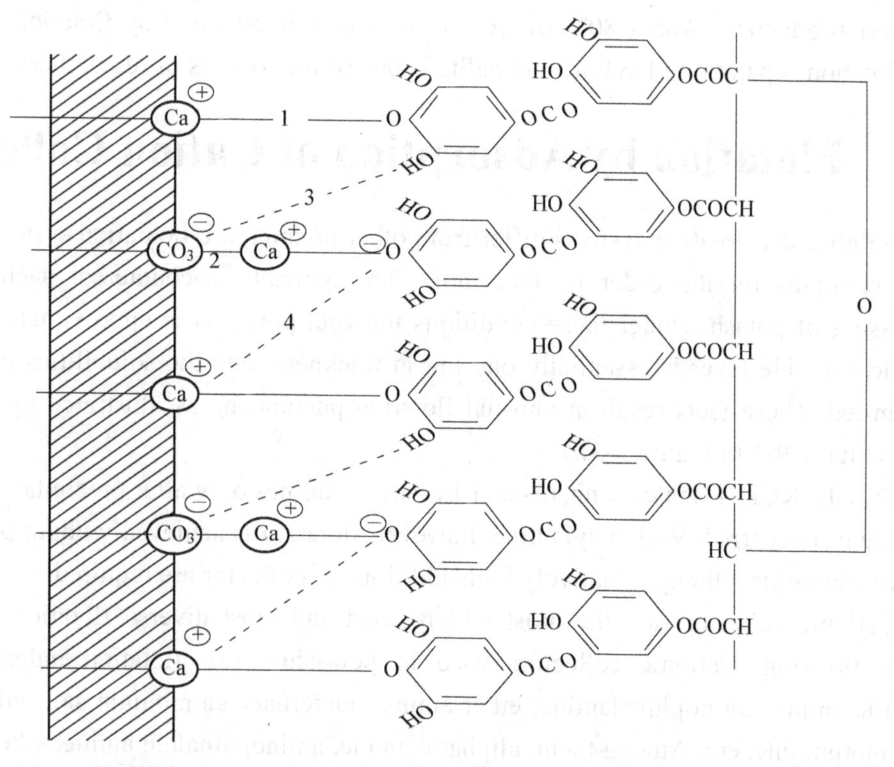

Figure 8-33 Adsorption of tannins on the surface of calcite

1—direct Ca—O bond; 2—Ca^{2+} active bond; 3—hydrogen bond; 4—electrostatic attraction

Chapter 9
Flotation Chemistry of Soluble Salt

Soluble potash mineral resources are the main raw materials for the production of potash fertilizer and other potash products, and play an extremely important role in industrial and agricultural production. Flotation is one of the main methods of potash fertilizer production. About 80% of KCl in the world is obtained by flotation production. The flotation separation of sylvite and halite is one of the focuses of researchers.

9.1 Flotation by Adsorption of Cation Collectors

Soluble salt flotation systems differ from other nonmetallic floatation systems in which ionic strengths on the order of five molar are typically encountered, such as in the processing of potash. Under these conditions the zeta potential is approximately zero, the electrical double layer is essentially one ion in thickness, and the solubilities of collectors are limited. These facts result in unusual flotation phenomena as illustrated by sylvite and halite (Figure 9-1 to Figure 9-3).

Sylvite, KCl, is floated with 12- and 14-carbon amines only after precipitation of amine chloride has occurred. With octylamine, however, flotation is achieved without precipitation of amine chloride although relatively high additions of collector are required.

Cationic collectors are the most widely used and most diverse flotation reagents in potash flotation. Cationic collectors used in potassium salt flotation industry include aliphatic amine, aminophthalamine, ether amine, quaternary ammonium salt, mixed amine, alkyl morpholine, etc. Among them, aliphatic amine, aminophthalein amine, ether amine and mixed amine are all used in the flotation of KCl and kainite ($KCl \cdot MgSO_4 \cdot 3H_2O$). In particular, octadecyl primary amine is the most widely used aliphatic amine. Most KCl fertilizers in the world use octadecyl primary amine for flotation production, and KCl in Qinghai and Xinjiang, the main producers of potassium salt in China, also uses octadecyl amine for flotation production.

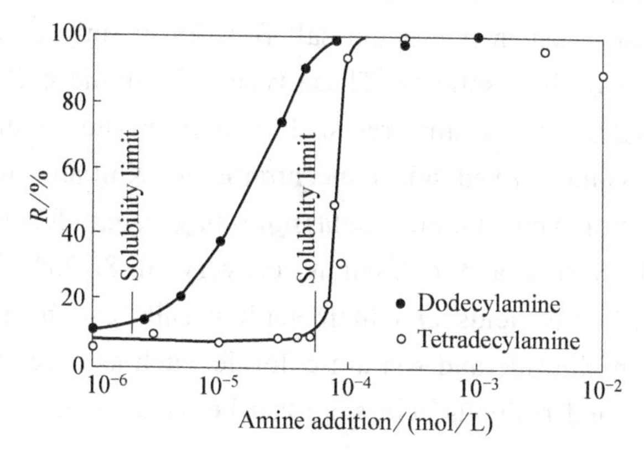

Figure 9-1 Flotation recovery of KCl as a function of amine addition

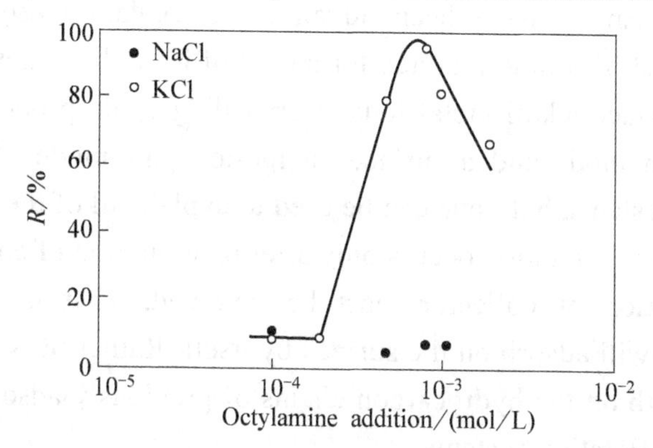

Figure 9-2 Flotation recovery of KCl and NaCl as a function of octylamine addition

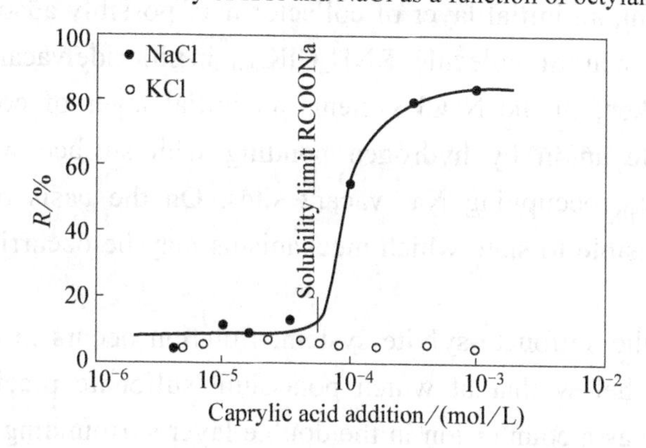

Figure 9-3 Flotation recovery of KCl and NaCl as a function of caprylic acid addition

9.2 Flotation by Adsorption of Anionic Collectors

In the case of halite, NaCl, flotation is not achieved with amines under any circumstances. On the other hand, good recovery is obtained with carboxylate collectors after the particular sodium carboxylate has precipitated. Sylvite does not respond to flotation with a 10-carbon carboxylate as collector, but complete flotation is effected with a 10-carbon sulfonate below the solubility limit of potassium decyl sulfonate.

Anionic collectors used in potassium salt flotation mainly include alkyl sulfonates, cycloalkyl sulfonates and alkyl sulfates. These types of anionic collectors can be used to flotation NaCl from sulfate potassium ores such as picromerite, aphthitalite and epsomite. For example, a potassium mixed-salt concentrator in Xinjiang used sodium dodecyl sulfonate as the collector, and obtained potassium-magnesium fertilizer concentrate with potassium content of 17.86% and potassium recovery of 82.73% by flotation. In fact, straight-chain saturated fatty acids and their sodium salts can be used for the flotation separation of potassium chloride and sodium chloride, such as *n*-octanoic acid, lauric acid, oleic acid, stearic acid and resinous acid soap can be used for the flotation of NaCl from sylvite salt ore.

A number of premises have been advanced to explain these phenomena: an ion exchange model, a heat of solution model, formation of insoluble reaction products between collector ions and surface alkali metal ions, a crystallographic properties model, collector and surface hydration mode and a surface charge-ion pair model. Each can be used to explain some of the systems, but none can be used to explain all of the observed behavior.

In those cases where flotation occurs only after precipitation of collector, phenomena in addition to precipitation of collector must be involved. That is, it is not likely that precipitated collector will adsorb on the surface by itself. Rather, it is more likely that these precipitates will adsorb on the hydrocarbon chains of previously adsorbed collector species as is the case in other flotation systems.

In the KCl system, an initial layer of collector may possibly adsorb as aminium ion in vacant K^+ sites or as a neutral molecule, $RNH_3ClK_{(aq)}$, in chloride vacant sites.

By the same token, in the NaCl systems an initial layer of collector may possibly adsorb as carboxylate anion by hydrogen bonding with surface water or as a neutral molecule, $RCOONa_{(aq)}$, occupying Na^+ vacant sites. On the basis of the available data, however, it is not possible to state which mechanisms may be occurring in initial collector adsorption.

With regard to the sulfonate-sylvite system, flotation occurs in a collector connector concentration region below that at which potassium sulfonate precipitates (Figure 9-4). Sulfonate ion adsorbs as a counter ion in the double layer surrounding the KCl by caprylate to the extensive hydration of carboxylate ion. This barrier of water molecules surrounding the carboxylate ion could prevent adsorption of collector on the surface.

In addition, the combination of amine collectors with different carbon chain lengths and different functional groups can not only obtain potash concentrate with higher quality and recovery, but also significantly reduce the dosage of reagents. Practice shows that whether anionic or cationic collectors are used, if they are mixed with an appropriate amount of hydrocarbon oil, they can often enhance the collection capacity of polar collectors, play a synergistic effect, improve the upper limit of floating particle size of minerals, and also reduce the dosage of polar collectors and reduce the cost of reagents.

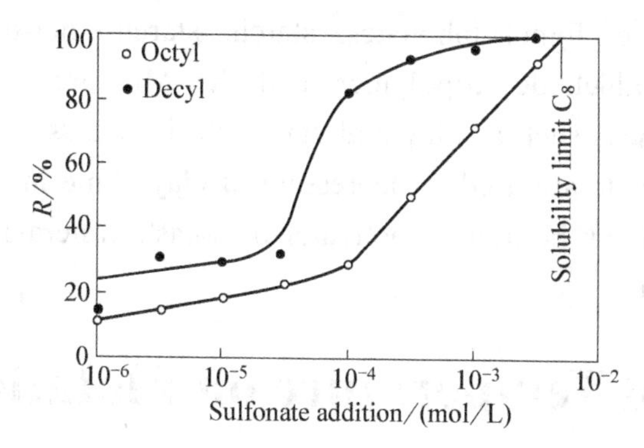

Figure 9-4 Flotation recovery of KCl as a function of octyl sulfonate addition

9.3 Modulation of Flotation

(1) Activation. Activators are rarely used in potassium salt flotation, because the surface of potassium salt minerals often works well with cationic or anionic collectors, and activators are basically not needed to improve the adsorption effect. Activators for potassium salt flotation include potassium ferrocyanide $K_4Fe(CN)_6$, xylene and lead nitrate. $K_4Fe(CN)_6$ and xylene are the surfactants of KCl when using fatty amine to flotation potassium salt, which can improve the KCl flotation index of concentrate. Lead nitrate is mainly used as an activator for NaCl and is used in flotation of NaCl from potassium salts with the anionic collector alkyl sulfate.

If lead and magnesium salts are added at appropriate pH values in the presence of carboxylates or sulfonates, flotation is effected. The results obtained with magnesium added as activator are presented in Table 9-1. At this concentration and pH, a significant amount of $MgOH^+$ would be present. With similar additions of lead at this pH, essentially all of the added lead would be $PbOH^+$. In these systems it seems likely that hydrogen bonding between surface water and the hydroxyl of the hydroxy complex occurs.

Table 9-1 Flotation response of NaCl from natural ore with nonanoic acid as collector and magnesium as activator

Collector addition	Activator addition	pH	Flotation response
2×10^{-4} mol/L	0.34 mol/L Mg^{2+}	5.5	No flotation
2×10^{-4} mol/L	0.34 mol/L Mg^{2+}	8.7	Flotation

(2) Depression. Potassium salt floatation depressant include phosphate, aluminum sulfate, tannin, amine aldehyde resin, carboxymethyl cellulose, lignosulphonates, starch,

starch-ether-based secondary amine condensate, urea-formaldehyde copolymer and KS-MF (a by-product of the condensation of urea and formaldehyde), etc. Phosphate and tannin can be used as depressant of kainite during flotation of sylvite-kainite. Amine aldehyde resin, carboxymethyl cellulose, lignosulphonates, starch, starch-ether-based secondary amine condensate, urea-formaldehyde copolymer and KS-MF are used as depressant of water-insoluble substances such as clay and carbonate in potassium salt flotation. Sodium and calcium lignosulfonate are used as depressant of clay slime in potassium salt flotation. In addition, humic acid can be used as activator of potash minerals such as soft potassium alum in potash flotation.

9.4 Effect of Temperature on Flotation

Because of solubility considerations, temperature displays a significant effect in these systems. These effects are shown in the flotation of a potash ore with two amines at three temperatures in Figure 9-5.

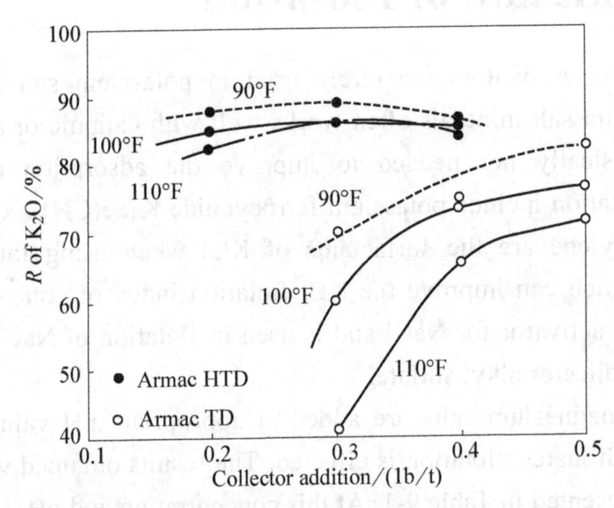

Figure 9-5 Flotation recovery of potash as a function of amine addition at elevated temperature

Comparison of the flotation data with the analytical data in Table 9-2 shows that the collection efficiency of the Armac TD which contains about 28% 16-carbon (saturated) amine, 25% 18-carbon (saturated) amine and a substantial amount of unsaturated amine decreases rapidly at temperatures above 90°F. The Armac HTD which contains about 25% 16-carbon (saturated) amine, 72% 18-carbon (saturated) amine and only a small amount of unsaturated amine is much less affected by the temperature changes. The necessity for precipitate formation in the KCl-amine system has been demonstrated as shown in Figure 9-1. The Armac HTD contains substantially more saturated octadecylamine, which will form precipitates of amine chloride at lower amine concentrations than will form with unsaturated amine species. The results obtained with each collector at 90, 100, and 110°F can be explained on a similar basis. The precipitates exhibit greater solubility at the higher temperature and the flotation response is accordingly less at the higher temperatures.

Table 9-2 Analytical data on amine flotation reagents used for temperature studies

Product classification		Armac TD	Armac HTD
		Tallow amine acetate	Hydrogenated tallow amine acetate
Amine acetate components	Tetradecyl/%	1	—
	Hexadecyl/%	28	25
	Octadecyl (saturated)/%	25	72
	Octadecenyl (mono-unsaturated)/%	46	3
Mean mol weight of primary amine acetate		322	323
Melting point (approximate)/℃		65	70
Average apparent molecular weight of the amine acetate		328	330

Note: Manufacturer's data.

Chapter 10 Slime Coatings and Carrier Flotation

10.1 Slime Coatings

The presence of slimes in flotation systems often results in deleterious effects on recovery and reagent consumption. Researchers have studied these phenomena extensively and have found that slime coatings are heaviest when the slime is uncharged or oppositely charged to the mineral being floated. The effect of ferric oxide slime on quartz flotation is presented in Figure 10-1. The pzc of quartz is pH 1.8 and that of the calcined Fe_2O_3 particles is pH 8. As the amount of slime is increase optimum flotation occurs only as both the Fe_2O_3 slime and quartz become negatively charged.

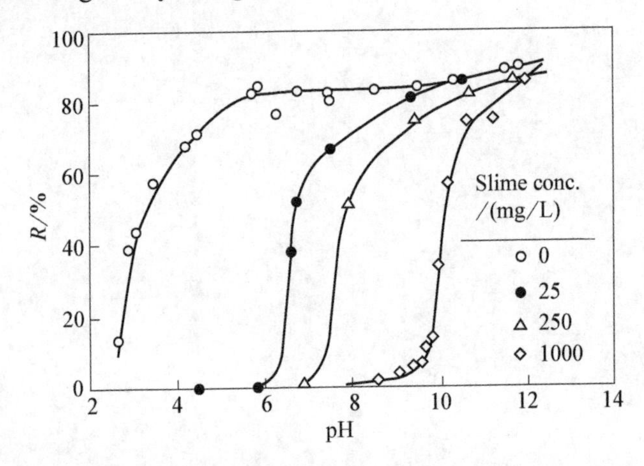

Figure 10-1 Effect of pH and ferric oxide slime on the flotation of quartz with 10^{-4} mol/L dodecylammonium acetate as collector

The effect of slimes on iron ore flotation was investigated. The results are presented in Table 10-1, and it can be noted that in soap flotation of calcium-activated quartz from natural iron ores, more iron oxide slime can be tolerated than is in amine flotation of quartz, probably because flotation is conducted at high pH where both the slime and quartz are negatively charged.

Similar phenomena have been observed in the galena-xanthate system in the presence of alumina slimes. See Figure 10-2. The pzc of alumina is approximately pH 9-10, and the positively-charged alumina particles form a slime coating on the negatively-charged galena particles.

Table10-1 Effect of slimes on the flotation of iron /%

Collector addition	Flotation products	Content			SiO_2 flotation recovery
		W_t	Fe	SiO_2	
0.3 lb/t dodecyl-ammonium chloride (deslimed), pH 6	Fe concentrate	20.4	59.6	10.6	
	SiO_2 froth	65.4	4.3	91.4	96.5
	Slime	14.2	25.2	59.5	
0.3 lb/t dodecyl-ammonium chloride (not deslimed), pH 6	Fe concentrate	84.9	19.0	71.4	
	SiO_2 froth	15.1	12.8	79.5	16.5
1.0 lb/t linoleic acid, pH 11 (deslimed)	Fe concentrate	21.1	60.6	9.4	
	SiO_2 froth	67.2	4.4	91.8	96.7
	Slime	11.7	27.9	54.1	
1.5 lb/t linoleic acid, pH 11 (not deslimed)	Fe concentrate	25.4	58.9	10.3	
	SiO_2 froth	74.6	4.5	92.6	96.4

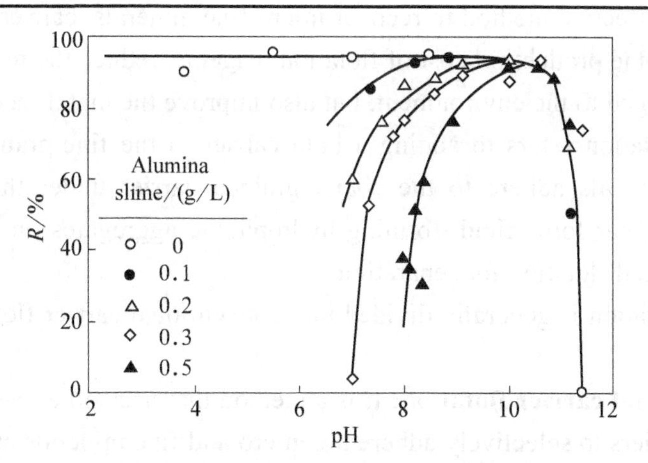

Figure 10-2 Percentage of flotation recovery vs pH for galena floated in the presence of various amounts of alumina slimes

The effect that slime exhibits on flotation response is also dependent on the particle size of the slime. It observed that goethite particles up to 23 microns in diameter inhibit quartz flotation with a cationic collector, whereas only goethite slimes finer than 5 microns affect anionic flotation of quartz. Also shown in Figure 10-3 is the fact that cationic flotation is much more sensitive to slimes than anionic flotation.

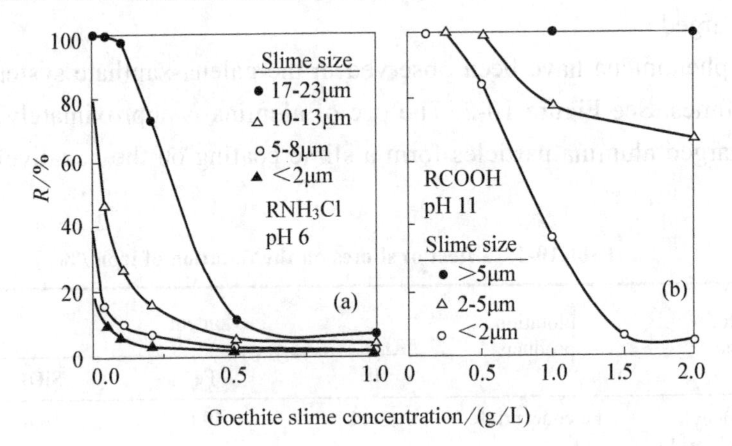

Figure 10-3 Effect of goethite slime size on the flotation of quartz

(a) with 10^{-4}mol/L dodecylammonium chloride as collector at pH 6 and

(b) with 10^{-4}mol/L linoleic acid activated with 50 mg/L calcium ion at pH 11

10.2 Carrier Flotation

With the increasing development of poor, fine and miscellaneous minerals and the requirements of concentrate quality, deep grinding will become the inevitable choice of beneficiation technology in the future, so the amount of secondary slime will also increase sharply. As an effective method to recover micro fine minerals, carrier flotation process can not only reduce the production cost of flotation reagents, reduce the total amount of residual reagents discharged to the environment, but also improve the metal recovery rate.

Carrier flotation refers to adding a bold carrier to the fine-grained pulp, making the fine-grained minerals adhere to the coarse-grained carrier under the action of selective surfactant and shear force field, forming hydrophobic aggregates on the surface, and then using conventional flotation for separation.

Carrier flotation is generally divided into conventional carrier flotation and self-carrier flotation.

Conventional carrier flotation: It is a method that uses other easy-floating coarse ore particles as carriers to selectively adhere the micro and fine minerals and float together with them. In the conventional carrier flotation process, the mineral used as the carrier is not the same mineral as the micro and fine minerals.

This process is mainly used to remove impurities in the materials. For example, some people abroad have carried out flotation test research on low-grade refractory lignite slime

(~0.038 mm) with strong hydrophobic coarse-grained coal as the carrier. The results show that the carrier particle size, fine slime and the content of carrier have important effects on the flotation process. Under suitable conditions, using strong hydrophobicity coarse-grained coal as carrier and isooctanol as flotation agent, clean coal with ash content of 8.30% and sulfur content of 0.72% can be obtained from lignite slime with ash content of 16.3% and sulfur content of 2.0%, and the recovery rate is 81.00%. The electrostatic attraction between carrier particles with little charge and fine particles lignite with high negative charge is an important reason for obtaining good flotation indexes.

Besides, some people use coarse-grained calcite as carrier to remove the fine-grained anatase of the clay, and use oleic acid as collector under the condition of pH 10.15, the residual amount of TiO_2 in the obtained clay concentrate is reduced from 72% to 39%, and the recovery rate of the clay is as high as 92%.

Self-carrier flotation: It is a flotation process using the same mineral as carrier. Compared with the conventional carrier flotation, it does not need to separate the carrier from the carried minerals. At the same time, the medicated coarse ore particles (generally flotation coarse concentrate) can better disperse and absorb, so as to reduce the total dosage of agentia and save the production cost.

For example, it shows that the self-carrier plays a significant role in the flotation of fine-grained ilmenite. When the proportion of coarse-grained carrier is more than 50%, good flotation effect can be achieved. The self-carrier flotation process was used to separate the refractory fine ilmenite in Panzhihua, and compared with the alone flotation process of fine minerals, the recovery rate of ilmenite with the size fraction of 0-20 μm increased from 52.56% to 61.96%.

Fankou Lead-Zinc Mine has realized the flotation recovery of fine-grained ore in the conventional flotation equipment by using the self-carrier flotation technology. The lead-zinc sulfide ore is added to the slime as carriers, the recovery rates of lead and zinc are increased by 6 and 12 percentage points respectively compared with the slime flotation alone. Besides, this technology has also achieved good results in the separation of fine-grained minerals such as sulfide oxidation mixed copper ore and tungsten ore.

References

[1] WANG Dianzhuo, HU Yuehua. Solution Chemistry of Flotation[M]. Changsha: Hunan Science & Technology Press, 1987.

[2] Fuerstenau M C, Miller J D, Kuhn M C. Chemistry of Flotation[M]. New York: Society of Mining Engineers of the American Institute of Mining, Metallurgical and Petroleum Engineers Inc, 1985.

[3] SUN Chuanyao, YIN Wanzhong. Flotation Principle of Silicate Mineral[M]. Beijing: Science Press, 2001.

[4] HU Yuehu, SUN Wei, WANG Dianzhuo. Electrochemistry of Flotation of Sulphide Minerals[M]. Beijing: Tsinghua University Press, 2009.

[5] NIE Guanghua. Selective depression and mechanism of fluorine-bearing minerals and calcium-bearing carbonate minerals[D]. Beijing: University of Science and Technology Beijing, 2015.

[6] Wills B A, Napier-Munn T J. Wills' Mineral Processing Technology[M]. Netherlands: Elsevier Science & Technology Books, 2006.

[7] XIE Guangyuan.Mineral Processing[M].Beijing: China University of Mining and Technology Press，2001.

[8] WANG Dianzhuo, QIU Guanzhou, HU Yuehua. Resource Processing[M]. Beijing: Science Press, 2010.